Event-based Model Predictive Control

IEEE Press

445 Hoes Lane

Piscataway, NJ 08854

Event-based Model Predictive Control

Bing Zhu

Beihang University
China

Zhigang Luo

AVIC Xi'an Flight Automatic Control Research Institute
China

Xiangyu Meng

Louisiana State University
USA

Zongyu Zuo

Beihang University
China

IEEE Press
WILEY

Published by John Wiley & Sons, Inc., Hoboken, New Jersey.

Library of Congress Cataloging-in-Publication Data

Names: Zhu, Bing, 1985- author | Luo, Zhigang, 1995- author | Zuo, Zongyu, 1982- author | Meng, Xiangyu, 1982- author
Title: Event-based model predictive control / Bing Zhu, Zhigang Luo, Xiangyu Meng, Zongyu Zuo.
Description: Hoboken, New Jersey : Wiley, [2026] | Includes bibliographical references.
Identifiers: LCCN 2026006454 | ISBN 9781394367788 hardback | ISBN 9781394367801 adobe pdf | ISBN 9781394367795 epub | ISBN 9781394367818 ebook
Subjects: LCSH: Predictive control—Mathematical models | Event detection—Mathematical models
Classification: LCC TJ217.6 .Z48 2026
LC record available at https://lccn.loc.gov/2026006454

Cover Design: Wiley
Cover Image: © Alexander Sikov/Getty Images

To Rui and Jiayi	*B.Z.*
To my parents	*Z.L.*
To Chunyan, Kaylee and Raelynn	*X.M.*
To Jingjing and Yanxi	*Z.Z.*

Contents

Foreword

Model predictive control (MPC) has become a cornerstone of modern control theory and practice, owing to its unique ability to handle multivariable dynamics, explicit constraints, and performance optimization within a unified framework. Over the past decades, MPC has found widespread applications in industrial processes, transportation systems, and aerospace engineering, while simultaneously stimulating substantial theoretical developments.

A persistent challenge in the implementation of MPC lies in the computational and communication burden associated with solving optimization problems at fixed and often dense sampling intervals. In resource-constrained or networked systems, such requirements can significantly limit practical applicability. Event-based control offers an effective alternative by updating control actions only when system-dependent conditions are satisfied, thereby reducing unnecessary computation and communication without sacrificing essential performance.

This book is devoted to the systematic study of event-based MPC. It presents a set of well-structured methodologies that integrate event-triggered and self-triggered mechanisms into MPC frameworks for both continuous-time and discrete-time nonlinear systems. Particular attention is paid to feasibility, stability, disturbance handling, and the avoidance of Zeno behavior. The authors further extend event-based MPC concepts to distributed and multi-agent systems, addressing communication efficiency and coordination challenges that are central to large-scale control applications.

A notable strength of the book lies in its balanced treatment of theoretical rigor and practical relevance. The proposed triggering strategies are carefully designed to reduce conservativeness while maintaining performance guarantees, and the analysis is supported by clear mathematical arguments and illustrative examples.

This book will be of interest to researchers, graduate students, and practitioners working in predictive control, event-based control, and networked or multi-agent systems. It serves both as a solid reference and as a meaningful contribution to the ongoing development of resource-efficient predictive control methodologies.

Xiaohua Xia, PhD
Professor, Fellow of IEEE
University of Pretoria

About the Authors

Bing Zhu received his BS and PhD degrees in Control Theory and Applications from Beihang University, Beijing, P.R. China, in 2007 and 2013, respectively. He was with University of Pretoria, Pretoria, South Africa, as a postdoctoral fellow supported by the Vice-Chancellor Postdoctoral Fellowship from 2013 to 2015. He was with Nanyang Technological University, Singapore, as a research fellow from 2015 to 2016. He joined Beihang University, Beijing, P.R. China as an Associate Professor in 2016. His research interests include model predictive control, smart sensing for UAVs and UGVs, and demand-side management for new energy systems. Dr Zhu serves as an Associate Editor for Acta Automatica Sinica.

Zhigang Luo received the BEng and MEng degrees in aircraft design from Beihang University (BUAA), Beijing, China, in 2017 and 2020, respectively. He obtained the PhD degree in control theory and control engineering from BUAA in 2025. Currently, he is an Engineer at Xi'an Flight Automatic Control Research Institute, Xi'an, China. His research focuses on flight control and optimization.

Xiangyu Meng received his BS degree in Information and Computing Science from Harbin Engineering University, China, in 2006, his MS degree in Control Science and Engineering from Harbin Institute of Technology, China, in 2008, and his PhD degree in Electrical and Computer Engineering from the University of Alberta, Canada, in 2014. From 2015 to 2016, he was a Research Fellow at Nanyang Technological University, Singapore, and from 2017 to 2018, he was a Postdoctoral Fellow at Boston University. He is currently an Associate Professor in the Division of Electrical and Computer Engineering at Louisiana State University, USA. His research interests include event-triggered control, model predictive control and reinforcement learning.

Zongyu Zuo received the BEng. degree in automatic control from the Central South University, Hunan, China, in 2005, and the PhD degree in control theory and applications from Beihang University (BUAA), Beijing, China, in 2011. Currently, he is a Full Professor with the School of Automation Science and Electrical Engineering, BUAA. He was an Academic Visitor with the School of Electrical and Electronic Engineering, University of Manchester, Manchester, UK, from 2014 to 2015, and held an Inviting Associate Professorship in mechanical engineering and computer science at UMR CNRS 8201, Université de Valenciennes et du Hainaut-Cambrésis, Valenciennes, France, in 2015 and 2017. His research interests include the fields of nonlinear system control, control of unmanned aerial vehicles, and coordination of multi-agent systems. He was identified as a Highly Cited Researcher by Clarivate Analytics. Prof. Zuo currently serves as an Associate Editor of IEEE Transactions on Industrial Informatics, *IEEE/CAA Journal of Automatica Sinica*, *Journal of the Franklin Institute*, *Journal of Vibration and Control*, and *International Journal of Aeronautical and Space Sciences*.

Preface

Model predictive control (MPC) is a numerical optimization-based control strategy that generates optimal control inputs by solving a constrained optimization problem over a finite prediction horizon. Traditional MPC methods require dense sampling, with optimization problems being resolved at every sampling instant, thereby imposing significant computational demands on processors. Event-triggered control, by contrast, leverages event-driven mechanisms and predefined thresholds, triggering updates only when specific conditions are met. This allows the system to maintain satisfactory performance while reducing redundant sampling and control actions. Integrating MPC with event-triggered control offers an effective means of reducing the frequency at which optimization problems must be solved. This not only conserves computational resources but also enhances the practical viability of the algorithm, highlighting its substantial research potential.

Nonetheless, current event-triggered MPC methodologies face several challenges. For instance, event-triggered MPC typically requires continuous monitoring to evaluate trigger conditions.

Self-triggered MPC methods, due to conservative state estimations, tend to result in shorter triggering intervals. In disturbance-aware event-triggered MPC, disturbance information is often underutilized. Moreover, relatively few studies have explored event-triggered MPC in the context of multi-agent systems, leaving room for further investigation.

This book concentrates on event-triggered MPC methodologies. Through an in-depth analysis of event-triggering mechanisms, it explores strategies aimed at reducing redundant sampling and the frequency of optimization problem-solving, while enabling efficient information exchange among agents. The following investigations have been conducted from these perspectives.

For disturbed nonlinear continuous-time systems, a two-phase predictive event-triggered MPC method is proposed. In the first phase, the time interval is estimated based on the worst-case effect of potential disturbances, with state measurements taken only at the end of this phase. In the second phase, system states are sampled periodically, and triggering conditions are monitored continuously until the conditions are satisfied, yielding the corresponding triggering instant. This two-phase event-triggered MPC framework generalizes the concept of triggering, with both periodic event-triggered and self-triggered strategies considered as special cases. During periodic sampling, the notion of a predictive safety bound is introduced to mitigate the potential failure of triggering conditions caused by discontinuous measurements. By carefully designing the sampling period and triggering threshold, the upper and lower bounds of triggering intervals can be determined, effectively preventing the Zeno phenomenon. Moreover, appropriate parameter tuning achieves a balance between monitoring cost and computational load.

An event-triggered nonperiodic interval sampling MPC method is proposed for disturbed and constrained discrete nonlinear systems. This algorithm designs the triggering threshold based on Lyapunov convergence properties and computes the required time interval to reach the threshold under worst-case disturbance by evaluating the deviation between the actual and predicted states. After this interval elapses, a new sample is taken, and using updated state feedback, a new interval is computed until the sampling interval reaches zero—marking the next triggering instant. Additionally, a prediction horizon update strategy is introduced, leveraging the predictive nature of MPC to alleviate computational burdens during each optimization process. Compared with self-triggered MPC approaches, the event-triggered nonperiodic sampling strategy reduces conservativeness in estimating triggering intervals, enlarges the intervals, and lowers the frequency of solving optimization problems.

A disturbance-compensated composite event-triggered MPC method is proposed for constrained discrete-time systems subject to slowly varying disturbances. The controller consists of two components: the disturbance compensation controller, which utilizes the slowly varying nature of the disturbance to perform estimation and compensation, and the model predictive controller, which optimizes a performance index to achieve optimal control behavior. An event-triggered mechanism is employed to reduce the frequency of solving the optimization problem. Additionally, a predictive triggering strategy is introduced to estimate the future state in advance, facilitating rational triggering decisions. Finally, the feasibility and stability of the proposed composite MPC framework with disturbance compensation are analyzed theoretically, and the effectiveness of the algorithm is validated through simulation experiments.

An event-triggered distributed MPC (DMPC) with periodic intermittent sampling is proposed for multi-agent vehicular formation control. Within the distributed framework, a virtual leader is constructed to serve as a reference, and each agent tracks its desired relative position with respect to the virtual leader.

Formation cost terms are incorporated into each agent's optimization objective, and fixed-formation maneuvers are achieved through inter-agent information exchange. All agents operate under a unified timescale framework, performing state sampling and communication at a constant period. By employing the event-triggered mechanism with periodic intermittent sampling, the approach effectively reduces communication overhead and the frequency of solving optimization problems, enabling streamlined information transmission while alleviating computational burden on processors.

October 2025

Bing Zhu and Zongyu Zuo
Beihang University

Zhigang Luo
AVIC Xi'an Flight Automatic Control Research Institute

Xiangyu Meng
Louisiana State University

Acknowledgments

The authors owe great thanks to Mr John Meng, Mrs Nandhini Karuppiah, Mr Aloysias Saint Thomas, and Ms Becky Cowan from Wiley for their valuable help and guidance during the publication process.

Acronyms

$\mathbb{N}_{\geq 0}$	Set of all nonnegative integers
$\mathbb{N}_{[p,q]}$	Set of all integers between p and q
$a \in W$	Element a belongs to set W
$\mathbb{X} \subseteq \mathbb{Y}$	Set $\mathbb{X}$ is a subset of Set $\mathbb{Y}$
$\lfloor a \rfloor$	The maximal integer no greater than a
$\lceil a \rceil$	The minimal integer no less than a
$\bar{\lambda}(A)$	The largest eigenvalue of matrix A
$\underline{\lambda}(A)$	The smallest eigenvalue of matrix A
$\|x\|_G$	Norm of vector x, $\|x\|_G = \sqrt{x^T G x}$, and G is a positive definite matrix
$\oplus$	The Minkowski sum of sets, $X \oplus Y = \{x + y \| x \in X, y \in Y\}$
$\ominus$	The Pontryagin difference of sets, $X \ominus Y = \{z \| z + y \in X, \ \forall y \in Y\}$
$\mathbb{Y}/\mathbb{X}$	$\mathbb{Y}/\mathbb{X} = \{s \| s \in \mathbb{Y}, \ s / \in \mathbb{X}\}$
A^+	Moore–Penrose Pseudoinverse of matrix A

1

Introduction

1.1 Background and Motivation

Model predictive control (MPC), also known as receding horizon control, is an optimization-based control strategy. The fundamental concept of MPC dates back to the 1960s [1], and since the 1980s, it has been successfully applied to industrial systems to address constrained multivariable control problems. From that point onward, MPC emerged as an advanced control methodology and gathered increasing interest from both academia and industry. The term "model predictive control" comprises three elements—model, predictive, and control—each conveying essential aspects of the method's underlying principles.

Model: In practical control system design, engineers aim to gain deeper insights into the characteristics of physical systems and to construct accurate dynamic models. A well-defined model plays a critical role in controller synthesis. For systems lacking an explicit model, identification techniques can be employed to derive suitable representations. Despite potential mismatches and uncertainties in the identified models, they still serve as a solid foundation for controller design.

Predictive: The ability to make forward-looking predictions based on current conditions—coupled with strategic planning to achieve predefined objectives—is an innate biological capability. In the MPC framework, this predictive ability constitutes a core mechanism that enables the method to realize its superior performance characteristics.

Control: Control actions in MPC are determined by solving a mathematically tractable constrained optimization problem. This process entails necessary analysis of key properties of the resulting control system.

MPC offers the following features. It facilitates systematic and flexible handling of state and input constraints. It employs an optimization-centric strategy that can deliver near-optimal performance. The algorithm exhibits inherent robustness, which can be further enhanced through robust MPC formulations to mitigate external disturbances, parameter uncertainties, and noise. It harnesses predictive capabilities to proactively improve system performance. These attributes collectively establish MPC as a promising paradigm for future optimization-based control strategies.

Despite the promising prospects of MPC, the method relies on numerical optimization to compute control inputs, imposing stringent requirements on computational efficiency. Currently, most embedded system processors exhibit limited computational capabilities, thereby restricting the large-scale deployment of MPC. Furthermore, with the advancement of wireless communication technologies, the transmission of information between sensors and controllers is increasingly facilitated via wireless networks. Traditional communication protocols typically adopt time-triggered periodic transmission schemes, possibly resulting in communication and computation resource usage exceeding actual system needs, further constraining MPC applications.

Event-triggered control addresses these limitations by introducing an event mechanism and threshold condition such that updates occur only when specific criteria are met. The core concept behind the event-triggered control is to update the control only when the system violates predefined performance specifications. This enables a more favorable trade-off between control performance and overall communication load compared to conventional periodic schemes.

The integration of event-triggered control strategies with MPC can effectively harness this mechanism to reduce communication capacity and avoid unnecessary resource utilization. Moreover, it lowers the frequency of optimization problem-solving, thereby alleviating computational demands [2]. Event-triggered MPC continues to be a highly active research area, particularly in the context of systems subject to external disturbances, packet loss in communication, and multi-agent cooperation—each scenario presenting compelling challenges and valuable research opportunities.

Control of multi-agent systems has become one of the prominent research topics in modern control theory. Within the framework of MPC, multi-agent coordination has achieved a number of representative advancements in recent years [3, 4]. However, the practical implementation of MPC in multi-agent systems remains limited due to the inherent complexity of inter-agent network communication and the computational demands of solving predictive control problems. Consequently, reducing the frequency of communication among agents as well as the individual computational burden associated with MPC updates presents a significant research challenge. Current investigations in this area are still in early stages. Event-triggered multi-agent MPC methods offer a compelling direction for further exploration. By selectively updating control actions only when specific triggering conditions are met, such approaches have the potential to improve scalability and efficiency while preserving desirable overall system performance.

This book is dedicated to exploring event-triggered MPC methodologies, with focused investigations into four main areas:

- Event-based two-phase MPC for nonlinear continuous-time systems.
- Event-based aperiodic sampling MPC for nonlinear discrete-time systems.
- Event-triggered MPC for disturbed systems.
- Periodic sampling-based event-triggered MPC for distributed systems.

Through a detailed examination of event-triggering mechanisms, robust handling of system uncertainties, and the study of inter-agent communication protocols, the objective is to reduce communication load and minimize the frequency of optimization computations. These efforts aim to enhance the practical value of event-triggered MPC and facilitate its wider application across engineering domains.

1.2 Model Predictive Control

Modern industrial control systems are influenced by a variety of factors. In addition to physical constraints such as actuator saturation, considerations of safety, process requirements, and economic performance must also be addressed. For instance, system parameters like pressure and temperature must be maintained within permissible limits. When dealing with constrained systems, the goal is not only to achieve stabilization but also to ensure enhanced performance. Moreover, industrial processes are often characterized by high-dimensional complexity, making the precise formulation of mathematical models exceedingly difficult. Due to external disturbances and parameter variations, uncertainties arise that render optimal control strategies—derived solely from idealized models—ineffective in practical applications. Such mismatches may even result in severe degradation of control performance. Consequently, the field of control engineering demands a new class of optimization-based methodologies that can simultaneously satisfy various system constraints, optimize performance indicators, and exhibit robustness against uncertainties and disturbances.

To address the aforementioned needs, MPC was proposed by researchers in the 1970s. Due to its capacity to handle complex multivariable control problems and its ability to enforce input, output, and state constraints in real time within an optimization framework, alongside its inherent disturbance-rejection capabilities, MPC has found extensive applications in practical industrial processes. Its application spans critical military and civilian sectors, including petroleum, chemical engineering, transportation, and aerospace [5–7]. In parallel, the theoretical development of MPC has attracted considerable attention from the research community [8, 9].

The underlying mechanism of MPC can be described as follows: At each sampling instant, the controller utilizes current state measurements to formulate and solve an optimization problem over a finite prediction horizon. The computed control sequence is then implemented by applying the first—or in some cases, a subset—of its elements to the plant. At the next sampling instant, the optimization problem is updated with new measurements and solved, and this process continues in a receding horizon fashion. MPC is particularly well-suited for addressing multivariable constrained optimization problems that Proportional-Integral-Derivative (PID) control struggles to manage. As a model-based control strategy, MPC is capable of handling nonlinear, time-varying, and time-delay systems subject to complex constraints. The MPC algorithm typically involves three main steps:

- Prediction: Estimating the future evolution of the system using a predictive model.
- Optimization: Solving a finite-horizon constrained optimization problem via numerical methods.
- Implementation: Implementing a portion of the optimal control sequence on the actual system.

Compared with other control methodologies, the primary characteristics of MPC include (but not limited to):

- Model-based Prediction: MPC relies on a model that describes the dynamic behavior of the controlled system. This model is used to forecast future system responses, hence the term "predictive model"—which is the reason MPC is considered a model-based control strategy. Within the MPC framework, the model serves as a tool to predict the system's future trajectory based on current states and anticipated future inputs. Notably, the formulation of the model itself is not restricted to any specific representation. To adjust future outputs toward desired reference trajectories, control inputs are optimized accordingly. The form in which the predictive model is expressed (e.g., state-space equations, transfer functions, or convolution representations such as step and impulse responses) does not affect this predictive process. In fact, the model may also be derived from data using machine learning or other artificial intelligence techniques, making it suitable for complex systems where first-principles modeling is infeasible.
- Receding Horizon Optimization: Due to limitations in computational capacity, the prediction horizon used in numerically solving the optimization problem cannot be indefinitely long. As a result, MPC algorithms typically employ a finite prediction horizon. At each sampling instant, MPC computes an optimal control sequence by solving a constrained optimization problem, and only the initial portion of the solution is implemented to the system. This

process is carried out iteratively in a receding horizon fashion, allowing the controller to continuously update and correct for external disturbances and model mismatches. By solving the optimization problem at each step using the latest system measurements, MPC adapts to real-time changes and environmental uncertainties. This dynamic updating mechanism enhances control performance and enables more effective realization of optimal control objectives.

- Integration of Feedforward and Feedback in the Control Structure: Feedforward refers to the use of a predictive model to predict system state evolution over a finite time horizon, thereby enabling proactive compensation in the computation of control inputs. For instance, during vehicle turning or obstacle avoidance, the predictive model facilitates early response actions, enabling the control inputs with a degree of foresight. Since each optimization problem is solved based on measurements of the current system states or outputs, the resulting optimal control input can essentially be regarded as a function of the present measurements. When the optimization is repeated at the next sampling instant using updated measurements, the newly computed control inputs naturally incorporate this feedback. Thus, the MPC algorithm inherently exhibits a feedback structure, even while leveraging model-based predictions for feedforward control.
- Explicit Constraint Handling: MPC enables the explicit incorporation of constraints within the optimization framework, including constraints on system states and control inputs. The constraints are due to hard constraints imposed by environmental or operational conditions, as well as constraints resulting from control capacity. Compared to other control strategies, MPC exhibits a distinct advantage in simultaneously solving for optimal control actions while effectively addressing constraint-related challenges. This capacity to manage complex constraint scenarios underscores its strong practical relevance and engineering applicability.

MPC is a numerically-based control algorithm that emerged from industrial process control. Predictive control originated in practical industrial production and has since evolved into various distinct forms. One category includes industrial predictive control methods that were developed through continuous exploration and innovation within industrial applications. These approaches often rely on step or impulse response models, with typical representatives including dynamic matrix control and model predictive heuristic control [10, 11]. Although widely used in practice, these methods lack formal guarantees of closed-loop stability from a theoretical point of view.

Another class comprises adaptive predictive control methods, rooted in adaptive control and minimum variance control. A major contribution in this area was the Generalized Predictive Control (GPC) algorithm proposed in the 1980s [12]. GPC facilitated the development of quantitative analysis tools, but its application to complex nonlinear systems remains challenging, hindering further theoretical breakthroughs. Both industrial and generalized predictive control methods are commonly referred to as classical predictive control, characterized by their emphasis on practical engineering requirements rather than rigorous theoretical analysis.

Beginning in the 1990s, a new phase of predictive control research emerged, focused on achieving formal stability guarantees. MPC began to incorporate concepts from optimal control, along with invariant set theory and Lyapunov-based approaches. This shift enabled the design of controllers that could meet both stability requirements and performance specifications within a principled framework. Since then, MPC theory has matured significantly and continues to evolve through integration with modern information technologies. It has gradually become a prominent and emerging direction within the control community, offering tremendous potential for future applications [13–15].

Consider a general deterministic discrete-time system described by

$$x_{k+1} = f(x_k, u_k),$$

where x_k and u_k represent the system's state and control input at time step k, respectively. Both the state and input are subject to constraints:

$$x_k \in \mathbb{X}, \quad u_k \in \mathbb{U},$$

where $\mathbb{X}$ and $\mathbb{U}$ are convex, compact sets that contain the origin. The origin is assumed to be an equilibrium point of the system, and the system states are assumed to be available in real time.

The control objective is to stabilize the system to the origin. At the core of MPC is the real-time solution of an optimal control problem over a finite time horizon, based on a given performance index. This yields a control sequence that optimizes the predicted behavior of the future system states, where the system dynamics are used to forecast the trajectory. The finite-horizon optimal control problem is formulated as follows:

$$\begin{aligned}
&\min_{u_k} J(x_k, u_k) = \sum_{i=0}^{N-1} L(x_{k+i|k}, u_{k+i|k}) + V_N(x_{k+N|k}) \\
\text{s.t.} \quad & x_{k+i+1|k} = f(x_{k+i|k}, u_{k+i|k}), \\
& x_{k|k} = x_k, \\
& x_{k+i|k} \in \mathbb{X}, \quad u_{k+i|k} \in \mathbb{U}, \\
& x_{k+N|k} \in X_f,
\end{aligned} \tag{1.1}$$

where N denotes the prediction horizon. The control input sequence is represented by u_k, and x_k is the system state at time step k. The notation $x_{k+i|k}$ indicates the predicted state at time $k+i$, computed at time k. The stage cost function $L(x_{k+i|k}, u_{k+i|k})$ captures the intermediate performance over the horizon and is typically expressed in a quadratic form involving the state and control input, reflecting their respective weightings throughout the prediction. The terminal cost $V_N(x_{k+N|k})$ is defined as a quadratic function of the terminal state. The terminal set X_f imposes constraints on the terminal state. By properly designing the terminal cost and terminal set, it is possible to enforce a monotonic decrease in the performance index, thereby contributing to the stability of the control algorithm.

For a continuous-time system given by

$$\dot{x}(t) = f(x(t), u(t)),$$

where the system state and control input satisfy the constraints $x(t) \in \mathbb{X}$ and $u(t) \in \mathbb{U}$, respectively, an optimal control problem is solved at time t_k. The total cost function to be minimized is designed by

$$J(x(t), u(t)) = \int_{t_k}^{t_k+T} L(x(t), u(t))\, dt + V_f(x(t_k + T)), \tag{1.2}$$

where T denotes the prediction horizon, and $u(t)$ is the predicted control signal over the interval. The stage cost function $L(x(t), u(t))$, typically expressed as a quadratic function of the state and input, captures their relative importance throughout the prediction horizon. The terminal cost $V_f(x(t_k + T))$ is usually a quadratic function of the terminal state, representing its weighted contribution at the end of the horizon. The total cost function combines the accumulated stage costs with the terminal cost, thereby accounting for both the cumulative influence of state and input over the prediction horizon and the desired behavior of the system at its terminal state.

In the stability analysis of MPC closed-loop system, the optimal cost function is commonly employed as a Lyapunov candidate. For discrete-time systems, let u_k^* denote the optimal control input sequence obtained at time step k, with its associated cost function denoted by J_k^*. Similarly, u_{k+1}^* and J_{k+1}^* represent the optimal control input and corresponding cost function at time $k+1$. Since these optimal control sequences are independently computed based on the system states at k and $k+1$, they lack a direct analytical relationship—making it challenging to rigorously establish the connection between successive Lyapunov candidate functions J_k^* and J_{k+1}^*. To address this issue, researchers introduced a feasible solution approach in stability analysis. By constructing a feasible (not necessarily optimal) control input sequence u_{k+1} at time $k+1$, and denoting its cost as J_{k+1}, an intermediate trajectory is established that facilitates a comparison between J_k^* and J_{k+1}^*.

Early literature often adopted the simplifying assumption of a zero terminal set, ensuring that the predicted state at the terminal step N converges to the origin. During theoretical analysis, the latter part of the feasible control sequence was frequently set to zero, which allowed derivation of asymptotic stability conditions relatively easily. However, the imposition of a zero terminal constraint—despite its analytical convenience—can render the optimization problem infeasible in practice, or restrict its feasibility set to an impractically small region.

Subsequent studies relaxed the terminal equality constraint and introduced the concept of terminal inequality constraints together with invariant set theory [16]. The terminal set is selected as a small neighborhood around the origin wherein a local feedback control law guarantees system stability. Crucially, this terminal set is designed as a control invariant set, meaning that under the specified feedback policy, the system state remains within the set indefinitely. This refinement allows the construction of a feasible control input sequence at each time step that connects J_k^* and J_{k+1}^*, thereby enabling rigorous demonstration of system stability [17]. Consequently, the integration of terminal cost, terminal set, and terminal control law concepts provides the foundation for a theoretically sound MPC framework, one that successfully addresses feasibility and stability concerns for constrained discrete-time systems [8].

1.3 MPC in the Presence of Uncertainties

One critical issue in MPC is how to effectively manage system uncertainties, such as additive disturbances and parameter uncertainties. Owing to its receding-horizon optimization structure, standard MPC exhibits a degree of inherent robustness against small uncertainties [18]. However, to ensure robust stability, constraint satisfaction, and favorable closed-loop performance under uncertainty, researchers have developed rigorous theoretical frameworks for robust MPC.

According to the existing literature, three main approaches have been proposed for addressing robustness in MPC:

- Tube-based Methods: These approaches construct a bounded tube around the nominal trajectory to contain all possible state deviations due to disturbances, ensuring constraint satisfaction and stability.
- Constraint-tightening Methods: By systematically contracting the feasible constraint sets, these methods proactively account for worst-case uncertainty effects during optimization.
- Min-max Optimization Strategies: These methods explicitly formulate a min-max optimization problem to derive control policies that perform optimally against the worst-case disturbance scenarios.

For systems subject to uncertainty, the tube-based MPC approach was proposed. By adopting simplified parameterized local control policies within the optimization, this method facilitates tractable implementation while partially sacrificing optimality [19]. Due to uncertainties, each control action induces a set of possible trajectories, collectively referred to as a tube, where each trajectory corresponds to a specific uncertainty realization. The core idea of tube-based MPC is to impose additional constraints on the optimization problem such that all possible trajectories within the tube remain confined to a small neighborhood around a nominal trajectory [20]. As a result, tube-based MPC preserves a computational complexity comparable to that of standard MPC, while achieving less conservative closed-loop performance.

The fundamental idea behind tube-based MPC is to tighten state constraints around a nominal trajectory using a local feedback control law, ensuring that all possible realizations of system uncertainty remain within the constraints. This is accomplished by constructing a positively invariant set that bounds the deviation between the actual and predicted system states under feedback control [21]. In Ref. [19], the state and input constraint sets are contracted to

$$\bar{X} = X \ominus S, \quad \bar{U} = U \ominus KS,$$

where K is a stabilizing feedback gain, S is a disturbance invariant set, and $\ominus$ denotes the Pontryagin difference between sets. This constraint tightening ensures that even in the presence of disturbances, the true system states and inputs stay within the original admissible sets.

Later work [22] introduced an improved formulation by including the nominal system's initial state as a decision variable in the MPC optimization problem. This allowed the feasible set to expand and enabled robust exponential stability. The key innovation is that the nominal initial state is optimized online rather than fixed. This design offers several advantages: the cost function is zero inside a robust invariant set Z, facilitating proofs of stability and attraction; the terminal set associated with the tube is no longer constrained to lie within Z, which relaxes restrictions compared to deterministic MPC formulations; it can be shown that the closed-loop system is robustly exponentially stable with respect to the set Z. This tube-based MPC framework has been extended to handle linear time-varying systems [23] and nonlinear systems [24], further broadening its applicability and robustness in real-world control scenarios.

In the aforementioned tube-based MPC methods, both the feedback gain and the positively invariant set are typically computed offline. Consequently, the cross-sections of the state and input tubes remain fixed throughout the optimization, yielding a framework often referred to as rigid tube MPC, which maintains computational complexity comparable to standard MPC. However, when the system uncertainty is state-dependent, the inability to adapt tube geometry online can lead to overly conservative performance. This limitation has driven the development of dynamic tube MPC strategies.

Dynamic tube MPC introduces tube parameters as decision variables within the optimization problem, allowing the tube's geometry, specifically its shape and size, to be updated online in response to system variations. Several representative approaches have emerged from this paradigm. In similarity-based tube MPC [25], the shape of the tube is predetermined offline, while its magnitude is adjusted during runtime. Elastic tube MPC [26] enhances this concept by treating the tube geometry as a vector-valued decision variable, thus enabling simultaneous optimization of shape and size. Furthermore, decoupled state-feedback tube MPC [27] parameterizes state predictions and control inputs using partial sequences, constructing tube as combinations of these subcomponents. This formulation facilitates finer control over tube geometry compared to elastic tube MPC, potentially yielding improved closed-loop performance.

Despite their benefits, dynamic tube MPC methods inevitably introduce additional decision variables and structural complexity into the optimization process. As such, their implementation requires a careful balance between computational burden and reduction in conservativeness to ensure practical applicability.

Tube-based MPC can be viewed as a feedback MPC strategy that incorporates constraint-tightening techniques. Its implementation relies on appropriately parameterizing the control input to shape bounded tubes around the nominal system trajectory. For linear systems, such parameterizations are relatively straightforward to construct. However, due to the inherent complexity of nonlinear dynamics, designing suitable tube structures in nonlinear systems is significantly more challenging.

As an alternative to tube-based approaches in nonlinear settings, researchers have proposed constraint-tightening methods within open-loop optimization [28]. For example, Ref. [29] employed Lipschitz constants to quantify the error between the actual and predicted states, using this information to contract the admissible state constraint set accordingly. In contrast, Ref. [30] avoided explicitly constructing error sets and instead imposed robustness constraints directly on the predicted states using Lipschitz bounds. Building on this framework, the authors developed a robust distributed MPC scheme, where predicted state trajectories are both monotonic and bounded. Although these constraint-tightening methods may yield more conservative performance than nonlinear tube-based MPC due to their open-loop nature, their optimization problems are simpler and more tractable. This computational advantage makes constraint-tightening-based nonlinear MPC methods particularly attractive for practical applications.

When both parametric uncertainty and additive disturbances are present, the design requirements for robust MPC become increasingly demanding. Min-max MPC addresses this by minimizing a worst-case cost function over all possible uncertainty realizations, differing from tube-based approaches by explicitly optimizing for the worst disturbance scenario. Initially proposed for Single-Input-Single-Output (SISO) systems [31], it was later extended to general linear [32] and Linear Parameter-varying (LPV) systems [33]. For nonlinear systems, open-loop min-max formulations [28] suffer from limited domains of attraction and computational burden. To mitigate conservativeness, feedback-based min-max MPC schemes incorporate state-dependent control laws [34], ensuring input-to-state stability [35]. Further improvements include frameworks that enforce explicit bounds on the closed-loop trajectories [36], and multistage MPC methods that approximate uncertainty via scenario trees, leveraging future measurements for enhanced robustness [37].

Despite its robustness benefits, the high computational complexity of min-max MPC limits its practical use. To address this, several methods have been developed to reduce computation. For example, Ref. [32] approximates worst-case scenarios using extreme disturbance realizations, effective only for low-dimensional systems. Offline strategies to approximate feedback min-max MPC solutions were proposed in Refs. [38] and [39]. In Ref. [40], convex reformulation via specific feedback design significantly improved efficiency. Hybrid schemes combining multistage MPC and tube-based MPC have also emerged [41], assigning large uncertainties to multistage optimization and smaller ones to tube methods. Furthermore, tailored algorithms have been developed for constrained nonlinear systems with network-induced delays and packet loss [42], actuator uncertainty [43], and to reduce communication costs via self-triggered control in general nonlinear systems [44].

Integrating disturbance observers (DOBs) with MPC enables simultaneous disturbance estimation and preemptive compensation, improving model accuracy and enhancing the robustness of MPC strategies. In Ref. [45], DOB-based compensation was applied to SISO affine nonlinear systems under MPC. Similarly, Ref. [46] proposed a DOB-augmented predictive torque control for induction motor systems, effectively mitigating parameter uncertainty and improving dynamic speed and torque response. Beyond direct compensation, estimated disturbance sequences can be embedded into the predictive model to reduce forecast errors. For example, Ref. [47] addressed slowly varying disturbances by generating predicted sequences from their rate of change, while Ref. [48] used statistical learning to model the relationship between disturbances, control inputs, and system states, enabling data-driven disturbance prediction.

1.4 Event-based MPC

In control systems, sampling, computation, and actuator updates are performed at fixed time intervals. This approach is known as time-triggered control. This methodology has been extensively studied over the past several decades. However, with the continued evolution of cyber-physical systems, communication networks have become central to information exchange among sensors, controllers, and actuators. Within this context, time-triggered control and communication strategies can significantly increase network load, potentially leading to congestion, delays, or packet losses that degrade system performance and, in some cases, induce instability.

To maintain satisfactory control performance while mitigating the risk of network congestion, it is essential to reduce communication overhead. Event-triggered strategies have proven to be an effective means of alleviating communication burden. An event is a specific, predefined condition that, when met, triggers a control-related action such as state sampling, control update computation, or data transmission. Within an event-triggered mechanism, the condition is typically defined as the moment when a calculated error function—which characterizes the deviation between the current system state and the state at the last update—exceeds a predefined threshold function. The core idea underlying event-triggered control and communication schemes is that information transmission and control input updates are executed only upon the occurrence of predefined events, while ensuring the desired control performance is maintained. Compared to time-triggered methods, event-triggered control and communication frameworks not only reduce the computational demand placed on the controller but also alleviate communication load across the network.

To effectively utilize limited computational and communication resources, data sampling, transmission actions, and control or estimation updates should be minimized while maintaining the desired control and estimation performance. This necessity has spurred substantial research into event-triggered mechanisms. An event-triggered mechanism typically comprises two key components.

- Error Function: This function characterizes the deviation between the current value and the value at the last triggering instant. Mathematically,

  $$f(e(t)) = f(x(t) - x(t_m))$$

  where t_m denotes the previous triggering time and t is the current time.
- Threshold Function: This function quantitatively defines the magnitude criterion.

Based on the interplay between the error function and the threshold function, the expression for determining the next event-triggering instant can be formulated by

$$t_{m+1} = \inf\{t > t_m \mid f(e(t)) \geq \sigma(t)\}. \tag{1.3}$$

By appropriately designing the triggering conditions, data sampling and control actions are executed only when necessary to ensure system stability and performance, thereby significantly reducing the consumption of available computational and communication resources. From the expression governing the triggering instants, two key insights emerge: (i) Threshold Function: This function directly influences the frequency of event triggering. Specifically, higher thresholds result in fewer triggering events, thus lowering the overall rate of event occurrences. Existing studies have explored a wide range of threshold functions to accommodate different system models and problem formulations. (ii) Performance–Resource Trade-off: A fundamental trade-off exists between the desired control performance and resource efficiency. As resource consumption decreases, overall control performance tends to deteriorate to some extent. This is primarily due to the reduced volume of data available for controller design and implementation. Therefore, it is essential to strike a balance that minimizes resource usage while preserving acceptable system performance.

Due to the distinctive properties of event-based control, the implementation of nonperiodic MPC based on event-triggering mechanisms has attracted increasing attention. Presently, event-based MPC approaches primarily consist of event-triggered MPC and self-triggered MPC schemes. In event-triggered MPC, the triggering mechanism determines the sampling and communication instants, thereby reducing both communication and computational loads. Event-triggered MPC activates control updates by continuously monitoring the deviation between the actual state trajectory and the optimal predicted trajectory, making it a promising research direction within event-triggered control strategies. Notably, given that MPC algorithms typically require considerable computational resources, embedding an event-triggering mechanism within MPC not only substantially reduces communication frequency but also alleviates the computational burden. The key advantage of event-triggered MPC lies in its capacity to lower the frequency of state sampling and optimization execution. The concept of event-triggered MPC was initially explored in [49], and has since garnered increasing research interest. In general, the existing literature on event-triggered MPC can be categorized into the following three groups:

1. Event-triggered MPC Based on Absolute Threshold Mechanisms: An implicit event-triggered MPC algorithm for nonlinear continuous-time systems was first proposed in [49]. The feasibility and convergence of the algorithm are ensured

provided that the inter-trigger intervals are bounded above and below. The trade-off between triggering frequency and control performance was examined in [50]. In [51], a constant-threshold event-triggering condition was introduced within a dual-mode MPC framework, which significantly reduced communication and computational power consumption compared to conventional periodic time-triggered MPC. Subsequently, [52] proposed a novel event-triggering condition for dual-mode MPC, where the threshold is formulated as a function of the minimum allowable inter-event time interval. In [53], input-to-state stability was analyzed for systems employing event-triggered mechanisms under quasi-infinite horizon MPC and dual-mode MPC frameworks.

2. Event-triggered MPC Based on Relative Threshold Mechanisms: By employing integral sliding mode control in conjunction with a relative-threshold-based event-triggering condition, a robust event-triggered MPC scheme was proposed in [54], capable of compensating for model uncertainties. In [55], an event-triggered nonlinear MPC framework was developed for distributed cooperative control of a group of dynamically decoupled nonlinear systems. Each subsystem is equipped with its own event-triggering condition, which is determined by the error derived from comparing local information with that of neighboring subsystems.

3. Event-triggered MPC Based on Lyapunov Functions: An event-triggered nonlinear MPC scheme based on Lyapunov functions was introduced in [56], where the triggering condition is derived by ensuring the monotonic decrease of the Lyapunov function starting from each self-triggered instant. In [57], two event-triggering conditions based on a non-monotonic Lyapunov approach were investigated, considering continuous-time nonlinear systems subjected to additive disturbances. In the context of distributed control, [58] studied a distributed event-triggered MPC framework for linear time-invariant systems, where cost-based event-triggering conditions are designed to determine when updated control data should be transmitted to other controllers, reducing inter-subsystem communication. The study in [59] proposed an event-triggering condition for distributed MPC that relies solely on local subsystem information. Later, [60] introduced neighbor-informed event-triggering conditions into distributed MPC. These conditions are derived from stability analysis, ensuring a continuous decrease in the Lyapunov function over successive triggering instants.

In the context of aperiodic MPC strategies, self-triggered MPC determines the next sampling instant solely based on the measured state obtained at the current triggering instant, without the need for continuous or periodic monitoring of the system state. As a result, state measurements are updated only at triggering instants, which significantly reduces communication burden. A widely adopted approach for designing self-triggered MPC strategies involves exploiting certain properties of the cost function to derive the triggering condition. For example, in [61], a self-triggered mechanism was developed based on the Lipschitz continuity of the cost function, along with guarantees of recursive feasibility and closed-loop stability. In [62], a sample-and-hold mechanism for control inputs was introduced between two consecutive triggering instants to further reduce communication demands. In [63], the researchers proposed a computationally efficient method that adaptively selects the sampling interval while ensuring a certain degree of performance sub-optimality. For systems with state constraints under disturbances, the original constraints can be appropriately tightened—essentially using the Gronwall–Bellman inequality—to ensure robust constraint satisfaction.

In addition, a class of self-triggered MPC methods introduces a new decision variable into the cost function, offering additional flexibility to balance communication cost and control performance. The basic idea is to relax the state penalty by adjusting this auxiliary variable when the controller is not triggered during the current interval. By solving a more complex optimization problem, the next triggering instant can be explicitly determined at the current triggering time. The reduction in the optimal cost can be interpreted as a reward for employing longer sampling intervals. Following this idea, [64] incorporated a term inversely proportional to the triggering interval into the cost function for an undisturbed linear system to model communication cost. Through optimization, a trade-off between communication load and control performance was effectively achieved. The work in [65] extended these results to linear systems subject to disturbances. To handle disturbances and enable output feedback, robust tube-based MPC and Luenberger observer-based approaches have been widely investigated. In [66], a disturbance-free nonlinear system was considered, and the standard cost function was modified by dividing the cost accumulated over the triggering interval by a constant greater than one. By treating the triggering interval as an additional decision variable, both the optimal control input and triggering interval can be determined simultaneously. For disturbed linear systems, [67] adopted the same strategy in combination with a tube-based MPC framework. In the case of linear systems with probabilistic constraints, [68] proposed a stochastic self-triggered MPC approach. For uncertain nonlinear systems, [44] developed a self-triggered control strategy based on min-max MPC and a similar relaxation mechanism.

Although most existing self-triggered MPC methods fall into the aforementioned two categories, there are a few notable exceptions. For instance, [69] considered continuous-time nonlinear systems and designed an MPC controller where the cost function over different time intervals and multiple prediction models are formulated in a discrete-time manner. Specifically, a zero-order hold mechanism is assumed to apply across all prediction intervals, and the holding time is treated as a function of the predicted system state. By solving an optimization problem, both the control input and the holding time can be simultaneously determined, resulting in a self-triggered controller. In [70], a relaxed dynamic programming inequality was incorporated into the MPC framework to construct a self-triggered strategy. By combining this strategy with a tube-based MPC approach, the authors investigated disturbed distributed linear systems.

1.5 Book Outline

The objective of this book is to integrate event-triggered mechanisms into MPC frameworks, and to explore event-triggered MPC strategies for both continuous-time and discrete-time systems. Moreover, it aims to incorporate disturbance attenuation into event-triggered MPC for uncertain systems, and to investigate the extension of event-triggered MPC approaches to multi-agent systems. The major challenges and key research difficulties involved include:

1. Design of Triggering Mechanisms: Absolute event-triggered mechanisms require continuous monitoring of the system state and real-time evaluation of the triggering condition by sensors, which imposes stringent demands on sensor hardware. Designing appropriate sampling and triggering mechanisms to reduce monitoring and decision frequency is a critical research issue.
2. Zeno Behavior in Continuous-time Systems: In event-triggered control of continuous systems, Zeno behavior may arise, referring to an infinite number of triggering events occurring in finite time—an undesirable phenomenon in practical systems. In event-triggered MPC, the system is triggered when the deviation between the actual and predicted states exceeds a threshold, meaning that the choice of triggering threshold directly influences the occurrence of Zeno behavior. How to select an appropriate threshold based on system characteristics, and how to compute the minimum inter-event time accordingly, remains a key design challenge.
3. Handling Bounded Disturbances: When the system is subject to bounded disturbances, how to effectively handle them becomes a crucial problem. One must design constraint-tightening schemes to ensure that the actual system behavior respects the imposed constraints. For systems with partial knowledge of the disturbances, leveraging such information in the design of the MPC scheme is also a significant research focus.
4. Event-triggered MPC in multi-agent Systems: The application of MPC in multi-agent systems requires information exchange among neighboring agents. Determining efficient and minimal communication protocols, and integrating them effectively with event-triggered MPC strategies, presents a complex and open challenge.

In response to the aforementioned research objectives and key challenges, the structure of this book and the relationships among its subsequent chapters are outlined as follows.

Chapter 2 presents an in-depth study of event-triggered mechanisms for continuous-time systems. A predictive two-phase event-triggered MPC approach is proposed, aiming to reduce the number of sampling instances and lower the triggering frequency.

Building on this, Chapter 3 addresses the discontinuous nature of states and integer-valued sampling intervals in discrete-time systems. By fully exploiting state information at sampling instants, an event-triggered aperiodic sampling MPC method is developed.

Furthermore, considering the slowly varying nature of disturbances in certain scenarios, Chapter 4 introduces a composite event-triggered MPC method with disturbance compensation.

Finally, in Chapter 5, the focus is extended from a single system to multiple disturbed systems. Taking into account the communication and interaction requirements among multiple systems, a distributed MPC method based on periodic sampling and event-triggered communication is proposed.

2

Two-phase Event-based MPC for Continuous-time Systems

Model predictive control (MPC) methods can handle optimal control problems with constraints. Traditional MPC methods typically use a very small sampling interval, requiring sensors to sample the state densely, while the controller needs to solve the optimization problem at each sampling moment, which places high demands on the processor's computational performance and communication load. Event-based MPC (EMPC) methods solve the optimization problem only when certain conditions are met, thus reducing the frequency of optimization problem-solving and alleviating the computational burden on the processor. Event-triggered MPC and self-triggered MPC are two main branches of EMPC. In event-triggered MPC, the system's state is continuously monitored, and the optimization problem is triggered based on whether the deviation between the current state and the desired state reaches a predefined threshold. In contrast, self-triggered MPC does not require real-time monitoring of the system state. Instead, it uses the system state obtained at the sampling moment to directly estimate the next triggering time and solves the optimization problem at that specific time. The main disadvantage of event-triggered methods is the need for specific sensors to continuously monitor the system state, as well as real-time determination of trigger conditions, which requires additional hardware. Self-triggered methods do not require specific sensor hardware, as they estimate the next trigger time using the sample information from the previous trigger moment. Due to the conservativeness of the estimation process, the trigger intervals are usually short, often leading to multiple triggers.

Based on the characteristics of both event-triggered and self-triggered methods, this chapter proposes a predictive two-phase event-triggered MPC method for nonlinear continuous systems subject to disturbances, referred to as the $\alpha - \beta$ strategy. In the first phase, the worst-case effect of disturbances is considered, and the time interval required to reach the trigger threshold is estimated, denoted as αT.

During this time interval, no sampling is performed, and state measurements are only taken after the first phase ends. After the first phase, the system enters the second phase, where periodic state measurements and checks are conducted at intervals of βT, until the trigger condition is met and the corresponding trigger time is obtained. In this chapter, a more general event-based MPC framework is proposed, where periodic event-triggered mechanisms and self-triggered mechanisms are special cases of this framework. The concept of a safety prediction value is introduced through predictive methods, ensuring the safety of interval sampling. By reasonably designing the trigger threshold and periodic sampling interval, upper and lower bounds for the trigger interval are derived, and the balance between computational cost and measurement cost can be achieved by adjusting the parameters.

2.1 Problem Statement

Consider the following continuous-time nonlinear system

$$\dot{x}(t) = f(x(t), u(t)) + w(t), \quad x(t) \in \mathcal{X}, \; u(t) \in \mathcal{U}, \; w(t) \in \mathcal{W}, \tag{2.1}$$

where $x(t) \in \mathcal{X} \subset \mathbb{R}^n$ and $u(t) \in \mathcal{U} \subset \mathbb{R}^m$ represent system states and controls; the set $\mathcal{X}$ and the set $\mathcal{U}$ are bounded closed sets, and contain the origin in their interior. Define the maximum norm of the constrained states by $b = \sup_{x\in\mathcal{X}} \|x\|_Q$. The disturbance $w(t) \in \mathcal{W} \subset \mathbb{R}^n$ is bounded, and its known bounds can be denoted by $\eta = \sup_{w(t)\in\mathcal{W}} |w(t)|$. The continuously differentiable function $f : \mathbb{R}^n \times \mathbb{R}^m \to \mathbb{R}^n$ satisfies $f(0,0) = 0$, and it is Lipschitz continuous with respect to x, i.e., $\|f(x_1, u) - f(x_2, u)\|_P \leq G \|x_1 - x_2\|_P$, where G is the Lipschitz constant. The matrix P is positive definite.

In case of uncertain disturbance, we have to use nominal model of (2.1) for prediction:

$$\dot{\hat{x}}(t) = f(\hat{x}(t), \hat{u}(t)), \tag{2.2}$$

where $\hat{x}$ and $\hat{u}$ are nominal state and control input, respectively.

Assumption 2.1 For Jacobian matrices $A = \partial f/\partial x|_{(0,0)}$ and $B = \partial f/\partial u|_{(0,0)}$, there exists feedback gain K such that all eigenvalues of $A_K = A + BK$ have negative real parts.

Based on Assumption 2.1, we have the following lemma.

Lemma 2.1 *[51] Suppose that 2.1 holds. Given positive definite matrices Q and R, there exists a constant ε, a feedback gain K and a positive definite matrix P, such that*

- *for the system* $\dot{x}(t) = f(x(t), Kx(t))$, *the set* $\Omega_\varepsilon = \left\{x(t) \,\middle|\, V(x(t)) = \|x(t)\|_P^2 \leq \varepsilon^2\right\}$ *is a control invariant set;*
- $\forall x(t) \in \Omega_\varepsilon$, *it holds that* $Kx(t) \in \mathcal{U}$, *and* $\dot{V}(x(t)) \leq -\|x(t)\|_{Q^*}^2$, *where* $Q^* = Q + K^T RK$.

Design process for ε, K and P in Lemma 2.1 can be found in [17]. With feedback control $u(t) = Kx(t)$, the set Ω_ε is actually a control invariant set for the nominal system.

The objective of this chapter is to design an event-based MPC algorithm framework for the nonlinear perturbed continuous system (2.1), ensuring that the system state converges to a small neighborhood of the origin while satisfying state and control constraints throughout the process. Additionally, the event-triggered mechanism is employed in the MPC to solve the optimization problem, thereby reducing the number of trigger detections and the computational frequency of the optimization solver.

2.2 Two-phase Event-triggered MPC Algorithm

2.2.1 Optimization Formulation

Let $\{t_k\}$, $k \in \mathbb{N}_{\geq 0}$ denote time instants when the optimization is solved. At time instant t_k, the cost function is defined by

$$\begin{aligned} J(x(t_k), \hat{u}(t\,|t_k)) = &\int_{t_k}^{t_k+T} \left(\|\hat{x}(t\,|t_k)\|_Q^2 + \|\hat{u}(t\,|t_k)\|_R^2\right) dt \\ &+\|\hat{x}(t_k+T\,|t_k)\|_P^2, \end{aligned} \tag{2.3}$$

where Q and R are positive weighting matrices; P is the terminal weighting matrix satisfying Lemma 2.1. The predictive nominal state series at t_k is given by $\hat{x}(t\,|t_k)$, and it satisfies $\hat{x}(t_k\,|t_k) = x(t_k)$. The predictive control series solved at t_k is denoted by $\hat{u}(t|t_k)$.

The optimization to be solved can be formulated as follows:

$$\hat{u}^*(t\,|t_k) = \arg \min_{\hat{u}(t|t_k)} J(x(t_k), \hat{u}(t\,|t_k)), \tag{2.4}$$

$$\text{s.t.}\ \ \dot{\hat{x}}(t\,|t_k) = f(\hat{x}(t\,|t_k), \hat{u}(t\,|t_k)),\ \ t \in [t_k, t_k+T], \tag{2.5}$$

$$\hat{u}(t\,|t_k) \in \mathcal{U},\ \ t \in [t_k, t_k+T], \tag{2.6}$$

$$\hat{x}(t\,|t_k) \in \mathcal{X} \ominus \mathcal{X}_{t-t_k},\ \ t \in [t_k, t_k+T], \tag{2.7}$$

$$\hat{x}(t_k+T\,|t_k) \in \Omega_{\varepsilon_f}, \tag{2.8}$$

where $\mathcal{X}_\tau = \left\{ x| \, \|x\|_P \leq \frac{\eta\sqrt{\bar{\lambda}(P)}}{G}(e^{G\tau} - 1) \right\}$ is the constraint-tightening set that takes consideration of disturbances. The set $\Omega_{\varepsilon_f} = \{x(t)| \|x(t)\|_P^2 \leq \varepsilon_f^2, 0 < \varepsilon_f < \varepsilon\}$ is the terminal constraint. The optimal predictive state series and control series are denoted by $\hat{u}^*(t\,|t_k)$ and $\hat{x}^*(t\,|t_k)$, respectively.

In formulating the above optimization problem, a state constraint-tightening method was employed to shrink the admissible state range. Since the optimization problem is solved for the nominal system without disturbances, the admissible nominal state set in (2.7) is obtained by subtracting the state constraint-tightening set $\mathcal{X}_{t-t_k}$ from the original state constraint set $\mathcal{X}$. After applying the optimal control sequence obtained from the optimization problem to the actual disturbed system, the true system state becomes the nominal state superimposed with the effect of disturbances. As the tightened state constraint set has already accounted for all possible disturbance effects and is incorporated as a constraint in (2.7), i.e., the nominal state set has been reduced by all possible disturbance-induced deviations, the actual system state can still satisfy the constraint set in the end.

Assumption 2.2 The set Ω_ε in Lemma 2.1 satisfies $\Omega_\varepsilon \subseteq \mathcal{X} \ominus \mathcal{X}_T$, where $\mathcal{X}_T$ is tightened constraint $\mathcal{X}_\tau$ at time $\tau = T$.

Assumption 2.3 At the initial time t_0, for $x_0 \in \mathcal{X}$, the optimization (2.4)–(2.8) in finite horizon is solvable.

Remark 2.1 Assumptions 2.2 and 2.3 are commonly used in the MPC literature [71] to theoretically ensure the feasibility of the optimization problem and the stability of the system.

Algorithms for solving nonlinear constrained optimization problems can be broadly categorized into two types. The first type includes numerical optimization algorithms, e.g., interior-point method, active-set method, and sequential quadratic programming (SQP) method. The second type comprises heuristic optimization algorithms, e.g., genetic algorithm, simulated annealing algorithm, and particle swarm optimization algorithm. Classical numerical optimization algorithms are capable of achieving fast convergence and stable solution quality, making them well-suited for problems with favorable properties such as convexity. In contrast, heuristic optimization algorithms are more suitable for handling complex problems with strong nonlinearities, but they generally require higher computational power. In this chapter, the SQP method is used to solve the optimization problem (2.4)–(2.8). Compared with heuristic algorithms, this iterative method requires lower computational resources and can achieve faster convergence of the solution.

After solving the optimization problem (2.4)–(2.8), the obtained optimal control trajectory $\hat{u}^*(t|t_k)$ is applied to the actual system (2.1). The control input is implemented over the time interval from the current time t_k to the next triggering time t_{k+1}, and is specifically given by:

$$u(t) = \hat{u}^*(t|t_k), \quad t \in [t_k, t_{k+1}). \tag{2.9}$$

At the next triggering time t_{k+1}, the optimization problem is solved again using the same procedure. Once the system state enters the terminal region Ω_ε, the control signal is switched to a state feedback controller Kx, and there is no longer a need to solve the optimization problem. This approach is known as the dual-mode control method.

2.2.2 Event-based $\alpha - \beta$ Strategy

Due to external disturbances, there will be a certain deviation between the actual state trajectory $x(t)$ and the predicted optimal state trajectory $\hat{x}^*(t\ |t_k\ )$. In absolute event-triggered strategies, the triggering condition is typically formulated as

$$\|x(t) - \hat{x}^*\ (t\ |t_k\ )\|_p = \sigma,$$

where σ denotes the triggering threshold. That is, a new triggering event is initiated when the deviation between the actual and predicted optimal state trajectories reaches the threshold σ, prompting the computation of a new optimization problem and the generation of a new control input trajectory. However, in absolute event-triggering mechanisms, it is necessary to monitor the actual system state in real time and continuously evaluate the triggering condition. This often requires dedicated hardware, imposes high demands on the performance of sensors and processors, and significantly increases computational cost.

To address this problem, this chapter proposes an $\alpha - \beta$ strategy based on interval sampling, which enables triggering with a finite number of state measurements. The desired trigger threshold is denoted by σ_d, and it can be calculated by

$$\sigma_d = \frac{\eta\sqrt{\bar{\lambda}(P)}}{G}(e^{\alpha TG} - 1) \tag{2.10}$$

where $\alpha \in (0, 1)$ is a predetermined parameter. The design of the expected trigger threshold σ_d is related to the perturbation upper bound and the first-phase sampling interval, with specific details discussed later. Due to the use of interval sampling, the trigger condition cannot be designed such that the deviation between the actual and predicted states is exactly equal to the expected trigger threshold; otherwise, there might be no solution. If the deviation between the actual and predicted states at the trigger instant is greater than the expected trigger threshold, system stability cannot be guaranteed. Therefore, when constructing the trigger condition, it needs to be modified to ensure that with interval sampling, the deviation between the actual and predicted states at the trigger instant is slightly less than or exactly equal to the expected trigger threshold. That is, it should satisfy:

$$\|x(t) - \hat{x}^*(t\,|t_k)\|_P \leq \sigma_d. \tag{2.11}$$

Inspired by both self-triggered and periodic triggering, this chapter proposes a more generalized $\alpha - \beta$ event-triggered strategy. The details of this triggering mechanism are elaborated below. Consider the time interval required to reach the desired trigger threshold under the worst-case impact of potential disturbances. Assuming the most recent trigger occurred at time t_k, the deviation between the actual state $x(t)$ and the predicted optimal state $\hat{x}^*(t|t_k)$ at time t satisfies the following, based on Lipschitz conditions and the Gronwall–Bellman inequality:

$$\begin{aligned}
&\|x(t) - \hat{x}^*(t\,|t_k)\|_P \\
&\leq \left\| x(t_k) + \int_{t_k}^{t} (f(x(\tau), \hat{u}^*(\tau)) + w(\tau))d\tau - \hat{x}(t_k\,|t_k) - \int_{t_k}^{t} f(\hat{x}(\tau), \hat{u}^*(\tau))d\tau \right\|_P \\
&\leq \int_{t_k}^{t} (G\,\|x(\tau) - \hat{x}^*(\tau\,|t_k)\|_P + \eta\sqrt{\bar{\lambda}(P)})d\tau \\
&\leq \frac{\eta\sqrt{\bar{\lambda}(P)}}{G}(e^{G(t-t_k)} - 1).
\end{aligned} \tag{2.12}$$

If the upper bound of the deviation equals the desired trigger threshold σ_d, i.e.,

$$\frac{\eta\sqrt{\bar{\lambda}(P)}}{G}(e^{G(t-t_k)} - 1) = \sigma_d,$$

the corresponding time can be calculated as $t_k + \alpha T$. This indicates that setting the trigger interval to αT is a sufficient condition to satisfy trigger expression (2.11). However, using αT as the trigger interval is overly conservative because it considers the worst-case impact of disturbances throughout the entire estimation process.

To reduce conservativeness while avoiding continuous sampling and trigger determination, we adopt a periodic sampling strategy after time $t_k + \alpha T$. The sampling period for this strategy is βT, where $\beta \in (0, \alpha)$ is an adjustable parameter. During periodic sampling, to prevent the state deviation at sampling instants from exceeding the desired trigger threshold and affecting system stability, we estimate the state deviation for the next sampling instant (after an interval of βT) in advance at each sampling time. Based on this estimated maximum deviation, we design a safety prejudgment value σ_2, which will be specified later. When the state deviation at a particular sampling instant is greater than the safety prejudgment value σ_2, it indicates that the next sampling might exceed the desired threshold, meaning the next sample would be unsafe. Consequently, this current sampling instant is then designated as the trigger moment. By employing the aforementioned $\alpha - \beta$ strategy, we can achieve more reasonable triggering while reducing conservativeness. The entire triggering process is illustrated in Figure 2.1.

When using the proposed $\alpha - \beta$ strategy, we must ensure that the deviation between the actual state and the predicted optimal state does not exceed the desired trigger threshold within any two consecutive sampling intervals. Therefore, to guarantee that the state deviation at the subsequent sampling time does not surpass the desired trigger threshold, the state deviation at the preceding sampling time must satisfy certain preconditions.

Specifically, at a sampling instant $t = t_k + \alpha T + m\beta T,\ m \in \mathbb{N}$, we proactively estimate the potential state deviation within the interval $[t, t + \beta T]$ under the influence of disturbances. This means at time t, we estimate the state deviation at time $t + \beta T$ considering disturbances. If we calculate that (2.11) holds at time $t + \beta T$ under any disturbance, it indicates that the state deviation will not exceed the desired trigger threshold σ_d within the interval $[t, t + \beta T]$. In this case, the next sampling can proceed at time $t + \beta T$. Conversely, if (2.11) does not hold at time $t + \beta T$, it suggests that the state deviation might exceed the desired trigger threshold σ_d within the interval $[t, t + \beta T]$, making the next sampling instant $t + \beta T$ unsafe. Consequently, the current sampling instant t is then designated as the trigger moment.

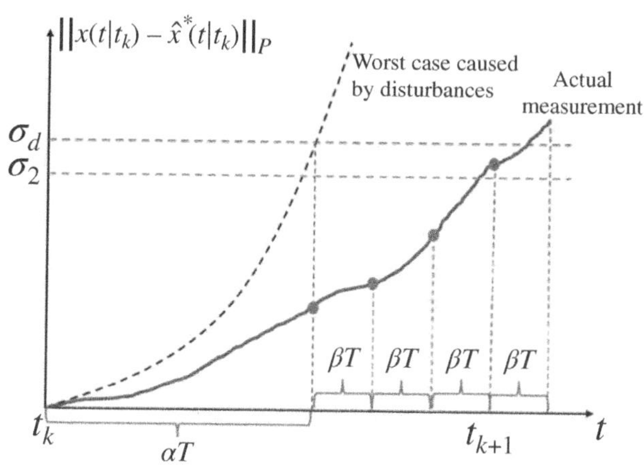

Figure 2.1 $\alpha - \beta$ event-based triggering strategy.

Let $x(t+\beta T|t)$ denote the state prediction at time t for time $t+\beta T$. According to the Gronwall–Bellman inequality, the deviation between the predicted state $x(t+\beta T|t)$ and the optimal state $\hat{x}^*(t+\beta T|t_k)$ satisfies the following inequality:

$$\|x(t+\beta T\,|t\ -\hat{x}^*\,(t+\beta T\,|t_k\,)\|_P \le e^{G\beta T}\,\|x(t)-\hat{x}^*\,(t\,|t_k\,)\|_P + \frac{\eta\sqrt{\bar{\lambda}(P)}}{G}\,(e^{G\beta T}-1). \tag{2.13}$$

To ensure (2.11) holds, the maximum deviation of $\|x(t+\beta T|t)-\hat{x}^*(t+\beta T|t_k)\|_P$ must not exceed the desired trigger threshold σ_d. Combining this with (2.13), it is clear that we only need the deviation between the actual state and the optimal predicted state at time t to satisfy:

$$\begin{aligned}\|x(t)-\hat{x}^*\,(t\,|t_k\,)\|_P &\le e^{-G\beta T}\left(\sigma_d-\frac{\eta\sqrt{\bar{\lambda}(P)}}{G}(e^{G\beta T}-1)\right)\\ &\le \frac{\eta\sqrt{\bar{\lambda}(P)}}{G}\,(e^{(\alpha-\beta)TG}-1)\triangleq\sigma_2.\end{aligned} \tag{2.14}$$

Define σ_2 to be the safety prejudgment value. It can be calculated by:

$$\sigma_2=\frac{\eta\sqrt{\bar{\lambda}(P)}}{G}\,(e^{(\alpha-\beta)TG}-1)\,.$$

The purpose of the safety prejudgment value σ_2 is to determine whether to proceed with the next sampling. If, at sampling instant t, the deviation between the actual state and the predicted optimal state satisfies (2.14), it signifies that the state deviation at the next sampling instant, $t+\beta T$, will not exceed the desired trigger threshold σ_d. This indicates that time $t+\beta T$ is safe, and sampling can occur then. Conversely, if (2.14) is not satisfied at sampling instant t, it implies that the state deviation might exceed the desired trigger threshold at the next sampling instant, $t+\beta T$. In this case, the next sampling is terminated, and the current sampling instant t is designated as the trigger moment.

Assuming the trigger moment is t_k, the condition (2.14) will be checked for the first time at $t_k+\alpha T$. Subsequently, it will be checked at intervals of βT until (2.14) no longer holds, where a new trigger moment t_{k+1} is determined.

To ensure the feasibility of the triggering algorithm, the interval between two consecutive triggers must not exceed the prediction horizon T, i.e.,

$$t_k+\alpha T+\gamma\beta T\le t_k+T.$$

Therefore, the maximum number of periodic samples should satisfy:

$$\gamma\le\gamma_{\max} \tag{2.15}$$

where $\gamma_{\max} = \lfloor (1-\alpha)/\beta \rfloor$, and $\lfloor \cdot \rfloor$ denotes the floor function. If condition (2.15) holds, then we can obtain:

$$t_{k+1} \leq t_k + \alpha T + \gamma_{\max}\beta T \leq t_k + T. \tag{2.16}$$

Therefore, as long as condition (2.15) holds, the interval between two triggers will not exceed the prediction horizon T, which guarantees the implementability of the algorithm.

Based on the analysis above, the entire $\alpha - \beta$ triggering strategy can be summarized as follows:

At a trigger moment t_k, solve the optimization problem (2.4)–(2.8) to obtain the optimal control trajectory $\hat{u}^*(t|t_k)$ over the interval $[t_k, t_k + T]$. Apply this control signal $\hat{u}^*(t|t_k)$ to the actual system within the interval $[t_k, t_k + \alpha T)$.

Then, at time $t_k + \alpha T$, measure the actual system state. Set $\gamma = 1$ and check both condition (2.14) and condition (2.15). If at least one of these conditions is not met, then the next trigger moment is determined to be $t_k + \alpha T$.

If both conditions are satisfied, then apply the control input $\hat{u}^*(t|t_k)$ during the interval $[t_k + \alpha T, t_k + \alpha T + \beta T)$. After this, system states are measured, the parameter γ is updated, and conditions (2.14) and (2.15) are evaluated at periodic intervals of βT. When either (2.14) or (2.15) is no longer satisfied, the most recent sampling moment is designated as the next trigger moment t_{k+1}.

Combining the above discussion, if the current trigger time is t_k, then the next trigger moment can be calculated by:

$$t_{k+1} = \min\{t|\ t \in S_{t_k}, \|x(t) - \hat{x}^*(t\ |t_k)\|_P > \sigma_2\}, \tag{2.17}$$

where

$$S_{t_k} = \{t_k + \alpha T + \gamma\beta T | 0 \leq \gamma \leq \lfloor (1-\alpha)/\beta \rfloor ; \gamma \in \mathbb{N}_{\geq 0}\}$$

represents the set of sampling moments;

$$\sigma_2 = \frac{\eta\sqrt{\bar{\lambda}(P)}}{G}\left(e^{(\alpha-\beta)TG} - 1\right)$$

denotes the safety prejudgment value; and $\beta \in (0, \alpha)$ is an adjustable parameter.

Lemma 2.2 *If an event-triggered $\alpha - \beta$ strategy is adopted, then the lower bound of the triggering interval is αT, and the upper bound of the triggering interval is $\alpha T + \gamma_{\max}\beta T$. In addition, condition (2.11) is always satisfied between any two triggering instants, and the maximum number of samplings between two adjacent triggering instants is $1 + \gamma_{\max}$.*

Proof: Suppose the previous trigger time is t_k. Then the first sampling time after t_k is $t_k + \alpha T$. Substituting $t = t_k + \alpha T$ into (2.12) yields:

$$\|x(t_k + \alpha T) - \hat{x}^*(t_k + \alpha T\ |t_k)\|_P \leq \sigma_d, \tag{2.18}$$

indicating that, at time $t_k + \alpha T$, the desired trigger threshold will not be exceeded. It also shows that αT is the lower bound of the trigger interval. We know that the trigger time t_{k+1} is determined by (2.17) and satisfies condition (2.15). When (2.15) holds, it can be derived from (2.16) that

$$t_{k+1} \leq t_k + \alpha T + \gamma_{\max}\beta T,$$

demonstrating that the upper bound of the trigger interval is $\alpha T + \gamma_{\max}\beta T$.

From (2.17), the next trigger time can be calculated by $t_{k+1} = t_k + \alpha T + \gamma\beta T$, where $0 \leq \gamma \leq \lfloor(1-\alpha)/\beta\rfloor$. If $\gamma = 0$, it is clear from (2.18) that the state deviation will not exceed the desired trigger threshold at time $t_k + \alpha T$.

Consider the case $\gamma > 0$. According to (2.17), we can deduce that:

$$\|x(t_{k+1} - \beta T) - \hat{x}^*(t_{k+1} - \beta T \,|t_k)\|_P \leq \sigma_2.$$

Substituting $t = t_{k+1} - \beta T$ into (2.13) yields:

$$\begin{aligned}
&\|x(t_{k+1}) - \hat{x}^*\,(t_{k+1}\,|t_k)\|_P \\
&\leq e^{\beta TG}\,\|x(t_{k+1} - \beta T) - \hat{x}^*(t_{k+1} - \beta T\,|t_k)\|_P + \frac{\eta\sqrt{\bar{\lambda}(P)}}{G}(e^{G\beta T} - 1) \\
&\leq e^{\beta TG}\sigma_2 + \frac{\eta\sqrt{\bar{\lambda}(P)}}{G}(e^{G\beta T} - 1) \\
&\leq \sigma_d
\end{aligned} \tag{2.19}$$

implying that the state deviation does not exceed the desired trigger threshold at t_{k+1}.

According to (2.17), the maximum number of samplings between two trigger instants is $1 + \lfloor(1-\alpha)/\beta\rfloor$. Combined with (2.15), it can be claimed that the maximum number of samplings between two triggers is $1 + \gamma_{\max}$. □

The two-phase event-triggered MPC algorithm can be summarized by Algorithm 1.

2.2.3 Generalization of the $\alpha - \beta$ Strategy

The $\alpha - \beta$ strategy proposed in this chapter can be extended to an $n\alpha - m\beta$ strategy, where n and m are adjustable design parameters. These parameters can be tuned based on the specific physical process and the different forms of disturbances present.

A strategy in the form of $n\alpha - m\beta$ can be considered as generalization of existing EMPC algorithms.

If $n = 0$ and $m = 1$, then the triggering strategy becomes a $0\alpha - \beta$ strategy. Under this strategy, the system state is sampled periodically at intervals of βT. This type of triggering mechanism falls into periodic event-triggered MPC [52]. If $n = 1$ and

---

Algorithm 1 Two-phase-based MPC Algorithm

Offline: Design the prediction horizon T, weighting matrices Q and R, and the periodic sampling parameter β. Calculate the desired trigger threshold σ_d Lipschitz constant G, feedback gain K, weighting matrix P, and set parameters ε and ε_f.

Online:

1: Initialization: $k = 0$, $t_k = 0$.

2: At time t_k, measure the system state $x(t_k)$. If $x(t_k) \in \Omega_\varepsilon$, proceed to step (6). Otherwise, solve the optimization problem formulated by (2.4)–(2.8).

3: During time interval $[t_k, t_k + \alpha T)$, implement $\hat{u}^*(t\,|t_k)$. Set $\gamma = 1$.

4: At the current time τ, measure the current state $x(\tau)$, check condition (2.14) and (2.15).

5: If both conditions (2.14) and (2.15) are satisfied, apply the control input $\hat{u}^*(t\,|t_k)$ to the actual system over the interval $[\tau, \tau + \beta T)$. Let $\gamma = \gamma + 1$, and return to step (4). Otherwise, set $t_{k+1} = \tau$, return to step (2), and set $k = k + 1$.

6: Implement $u = Kx$ as dual-mode control.

$m = 0$, then the triggering strategy is called an $\alpha - 0\beta$ strategy. Here, the triggering interval is designed as αT to satisfy feasibility and stability conditions. This triggering mechanism can be seen as a form of self-triggered MPC [2].

2.3 Feasibility and Stability

When using MPC, it is crucial to analyze optimization feasibility and closed-loop stability. Recursive feasibility, often simply called feasibility in MPC literature, refers to the existence of a solution for the optimization problem (2.4)–(2.8) at the next computation time t_{k+1}, given that a solution exists at the current computation time t_k. Closed-loop stability, on the other hand, investigates the stability of the closed-loop system when using the MPC method. This section will elaborate on these two important issues.

2.3.1 Recursive Feasibility of Optimization

Theorem 2.1 *Suppose that Assumptions 2.1–2.3 hold. The optimization problem (2.4)–(2.8) is recursively feasible, if the disturbances and terminal constraints satisfy the following conditions:*

$$\eta \leq \frac{(\varepsilon - \varepsilon_f)G}{(e^{TG} - e^{(1-\alpha)TG})\sqrt{\bar{\lambda}(P)}}, \tag{2.20}$$

$$\varepsilon_f^2 \geq \varepsilon^2 e^{-\frac{\lambda(Q^*)}{\bar{\lambda}(P)}\alpha T}. \tag{2.21}$$

Proof: Based on Assumption 2.3, suppose that the optimization problem is feasible at time t_0.

If the optimization problem (2.4)–(2.8) is feasible at time t_k, then at time t_{k+1}, we can construct a feasible control trajectory $\tilde{u}(t \mid t_{k+1})$ and its corresponding state trajectory $\tilde{x}(t \mid t_{k+1})$.

If $t_{k+1} < t_k + T$, the feasible control trajectory $\tilde{u}(t \mid t_{k+1})$ can be constructed as:

$$\tilde{u}(t|t_{k+1}) = \begin{cases} \hat{u}^*(t|t_k), & t \in [t_{k+1}, t_k + T), \\ K\tilde{x}(t|t_{k+1}), & t \in [t_k + T, t_{k+1} + T) \end{cases} \tag{2.22}$$

where the feasible state trajectory $\tilde{x}(t \mid t_{k+1})$ satisfies the nominal system (2.2), and $\tilde{x}(t_{k+1} \mid t_{k+1}) = x(t_{k+1})$. If $t_{k+1} = t_k + T$, then the feasible control $\tilde{u}(t \mid t_{k+1})$ is constructed as $\tilde{u}(t \mid t_{k+1}) = K\tilde{x}(t \mid t_{k+1})$, $t \in [t_k + T, t_{k+1} + T)$, which is a special case of (2.22).

To prove recursive feasibility, we need to demonstrate that the terminal constraints, state constraints, and input constraints in the optimization problem (2.4)–(2.8) can all be satisfied simultaneously when applying the control input $\tilde{u}(t \mid t_{k+1})$ over the interval $[t_{k+1}, t_{k+1} + T]$.

To begin, let us establish the context of the problem. We're looking at the interval $t \in [t_{k+1}, t_k + T]$. Our goal is to analyze the deviation between the predicted state $\tilde{x}(t \mid t_{k+1})$ and the optimal state $\hat{x}^*(t \mid t_k)$. This analysis relies on two key aspects: (i) the Lipschitz property of the nonlinear function $f(x, u)$; and (ii) the condition that the trigger time satisfies $\|x(t_{k+1}) - \hat{x}^*(t_{k+1} \mid t_k)\|_P \leq \sigma_d$. Given these, we can state that the deviation between the predicted and optimal states satisfies:

$$\begin{aligned}
&\|\tilde{x}(t \mid t_{k+1}) - \hat{x}^*(t \mid t_k)\|_P \\
&\leq \|\tilde{x}(t_{k+1} \mid t_{k+1}) - \hat{x}^*(t_{k+1} \mid t_k)\|_P + G\int_{t_{k+1}}^{t} \|\tilde{x}(\tau \mid t_{k+1}) - \hat{x}^*(\tau \mid t_k)\|_P \, d\tau \\
&\leq \sigma_d + G\int_{t_{k+1}}^{t} \|\tilde{x}(\tau \mid t_{k+1}) - \hat{x}^*(\tau \mid t_k)\|_P \, d\tau.
\end{aligned} \tag{2.23}$$

With Gronwall–Bellman inequality, (2.23) satisfies:

$$\|\tilde{x}(t \mid t_{k+1}) - \hat{x}^*(t \mid t_k)\|_P \leq \sigma_d e^{G(t - t_{k+1})}. \tag{2.24}$$

Based on the event-triggering mechanism in Algorithm 1, the following holds true between any two trigger instants:

$$t_{k+1} - t_k \geq \alpha T. \tag{2.25}$$

Substituting $t = t_k + T$ into (2.24), and combining it with (2.25) and the expected threshold (2.10), we obtain:

$$\begin{aligned} \|\tilde{x}(t_k + T \,|t_{k+1}\|_P &\leq \|\hat{x}^*(t_k + T \,|t_k\|_P + \sigma_d e^{G(t_k+T-t_{k+1})} \\ &\leq \varepsilon_f + \frac{\eta\sqrt{\bar{\lambda}(P)}}{G}(e^{TG} - e^{(1-\alpha)TG}), \end{aligned} \tag{2.26}$$

where, in the second inequality, the terminal constraint condition (2.8) is used. Substituting the disturbance condition (2.20) into (2.26) yields:

$$\|\tilde{x}(t_k + T \,|t_{k+1})\|_P \leq \varepsilon, \tag{2.27}$$

indicating that, by applying the control input $\tilde{u}(t \mid t_{k+1})$, the predicted system state $\tilde{x}(t_k + T \mid t_{k+1})$ will enter the set Ω_ε at time $t_k + T$.

Consider the state evolution over the interval $t \in [t_k + T, t_{k+1} + T]$. According to Lemma 2.1, we have:

$$\dot{V}(\tilde{x}(t \,|t_{k+1})) \leq -\|\tilde{x}(t \,|t_{k+1})\|_{Q^*}^2 \leq -\frac{\underline{\lambda}(Q^*)}{\bar{\lambda}(P)} V(\tilde{x}(t \,|t_{k+1})), \tag{2.28}$$

where $V(x) \triangleq \|x\|_P^2$. Applying the comparison lemma to (2.28) yields:

$$V(\tilde{x}(t \,|t_{k+1})) \leq \varepsilon^2 e^{-\frac{\underline{\lambda}(Q^*)}{\bar{\lambda}(P)}(t-t_k-T)}, \tag{2.29}$$

where $t \in [t_k + T, t_{k+1} + T]$. Using (2.25) and the terminal condition (2.21), substituting $t = t_{k+1} + T$ into (2.29) yields:

$$V(\tilde{x}(t_{k+1} + T \,|t_{k+1})) \leq \varepsilon^2 e^{-\frac{\underline{\lambda}(Q^*)}{\bar{\lambda}(P)}\alpha T} \leq \varepsilon_f^2. \tag{2.30}$$

Consequently, from (2.30), it follows that:

$$\|\tilde{x}(t_{k+1} + T \mid t_{k+1})\|_P \leq \varepsilon_f,$$

satisfying the terminal constraint condition.

Next, we analyze the feasibility of the state constraints. Over the interval $[t_{k+1}, t_k + T]$, since the optimization problem is feasible at time t_k, the following condition is satisfied:

$$\hat{x}^*(t \,|t_k) \in \mathcal{X} \ominus \mathcal{X}_{t-t_k}. \tag{2.31}$$

It follows from (2.24) that the relationship between the predicted state and the optimal state is given by:

$$\|\tilde{x}(t \mid t_{k+1})\|_P \leq \|\hat{x}^*(t \mid t_k)\|_P + \sigma_d e^{G(t-t_{k+1})}. \tag{2.32}$$

Define the set $E_\tau = \{x \mid \|x\|_P \leq \sigma_d e^{\tau G}\}$. Then, from (2.31) and (2.32), it follows that the predicted state satisfies:

$$\tilde{x}(t \mid t_{k+1}) \in \mathcal{X} \ominus \mathcal{X}_{t-t_k} \oplus E_{t-t_{k+1}}. \tag{2.33}$$

In another aspect, it follows from (2.25) that:

$$\begin{aligned} e^{G(t-t_k)} - 1 - (e^{\alpha TG} - 1)e^{G(t-t_{k+1})} &\geq e^{G(t-t_k)} - 1 - (e^{G(t_{k+1}-t_k)} - 1)e^{G(t-t_{k+1})} \\ &\geq e^{G(t-t_{k+1})} - 1. \end{aligned} \tag{2.34}$$

Multiplying both sides of (2.34) by $\frac{\eta\sqrt{\bar{\lambda}(P)}}{G}$ yields the following relation between the sets:

$$\mathcal{X}_{t-t_k} \ominus E_{t-t_{k+1}} \supseteq \mathcal{X}_{t-t_{k+1}}. \tag{2.35}$$

It then follows from (2.33) and (2.35) that:

$$\tilde{x}(t \mid t_{k+1}) \in \mathcal{X} \ominus \mathcal{X}_{t-t_{k+1}}, \tag{2.36}$$

implying that state constraints are satisfied for $t \in [t_{k+1}, t_k + T]$.

From (2.29), it is known that for $t \in [t_k + T, t_{k+1} + T]$, the predicted state $\tilde{x}(t \mid t_{k+1})$ lies within the set Ω_ε. Based on Assumption 2.2, it holds that

$$\tilde{x}(t \mid t_{k+1}) \in \Omega_\varepsilon \subseteq \mathcal{X} \ominus \mathcal{X}_T \subseteq \mathcal{X} \ominus \mathcal{X}_{t-t_{k+1}}. \tag{2.37}$$

As a consequence, from (2.36) and (2.37), it follows that the state constraints are satisfied over the interval $t \in [t_k + T, t_{k+1} + T]$.

Finally, we analyze the feasibility of the control input constraints. Since the optimization problem at time t_k is feasible and solvable, it follows that $\hat{u}^*(t \mid t_k) \in \mathcal{U}$ over the interval $[t_{k+1}, t_k + T)$. According to Lemma 2.1, we have $K\tilde{x}(t \mid t_{k+1}) \in \mathcal{U}$ over the interval $[t_k + T, t_{k+1} + T]$. Therefore, the control input constructed in (2.22) satisfies the control constraints over the entire interval $[t_{k+1}, t_{k+1} + T]$.

Based on the above analysis, when using the control input constructed in (2.22), the terminal constraint, state constraints, and control input constraints are all satisfied. Therefore, the optimization problem is recursively feasible. □

2.3.2 Stability of the Closed-loop System

Theorem 2.2 *If the actual system satisfies Assumptions 2.1–2.3 and the conditions in Theorem 2.1, and additionally meets the condition:*

$$\sigma_d e^{(1-\alpha)TG}(\varepsilon + \varepsilon_f) + \frac{2b\sigma_d}{G}(e^{(1-\alpha)TG} - 1) < \frac{\underline{\lambda}(Q)}{\bar{\lambda}(P)}\alpha T \varepsilon_f^2, \tag{2.38}$$

then, the system state will converge to the control invariant set Ω_ε within finite time. Furthermore, if the state enters the set Ω_ε at a sampling instant, the state feedback control law $Kx(t)$ is then employed. From that time onward, the system is input-to-state stable (ISS) with respect to the disturbance $w(t)$.

Proof: First, we prove that when $x(t) \notin \Omega_\varepsilon$, the system state will enter the set Ω_ε within a finite time.

Define

$$J^*(x(t_k)) = J(x(t_k), \hat{u}^*(t \mid t_k)) = \min_{\hat{u}(t|t_k)} J(x(t_k), \hat{u}(t \mid t_k)),$$

with $k \in \mathbb{N}_{\geq 0}$, as a Lyapunov candidate function, where the control input is $\hat{u}^*(t \mid t_k)$. Let the cost function under the feasible control input $\tilde{u}(t \mid t_{k+1})$ applied at time $x(t_{k+1})$ be denoted as $J(x(t_{k+1}), \tilde{u}(t \mid t_{k+1}))$. The optimal cost at time $x(t_{k+1})$ satisfies

$$J^*(x(t_{k+1})) \leq J(x(t_{k+1}), \tilde{u}(t \mid t_{k+1})).$$

Therefore, the difference between $J^*(x(t_{k+1}))$ and $J^*(x(t_k))$ satisfies:

$$J^*(x(t_{k+1})) - J^*(x(t_k)) \leq J(x(t_{k+1}), \tilde{u}(t \mid t_{k+1})) - J^*(x(t_k)). \tag{2.39}$$

Define

$$J(x(t_{k+1}), \tilde{u}(t \mid t_{k+1})) - J^*(x(t_k)) \triangleq \Delta V_1 + \Delta V_2 + \Delta V_3,$$

where

$$\Delta V_1 = \int_{t_k+T}^{t_{k+1}+T} \|\tilde{x}(t \mid t_{k+1})\|^2_{Q+K^TRK} dt + \|\tilde{x}(t_{k+1} + T \mid t_{k+1})\|_P^2 - \|\hat{x}^*(t_k + T \mid t_k)\|_P^2,$$

$$\Delta V_2 = \int_{t_{k+1}}^{t_k+T} \|\tilde{x}(t \mid t_{k+1})\|_Q^2 - \|\hat{x}^*(t \mid t_k)\|_Q^2 \, dt,$$

$$\Delta V_3 = -\int_{t_k}^{t_{k+1}} (\|\hat{x}^*(t \mid t_k)\|_Q^2 + \|\hat{u}^*(t \mid t_k)\|_R^2) \, dt.$$

First, for $t \in [t_k + T, t_{k+1} + T]$, it follows from Lemma 2.1 that:

$$\|\tilde{x}(t \mid t_{k+1})\|^2_{Q+K^TRK} = \|\tilde{x}(t \mid t_{k+1})\|^2_{Q^*} \leq -\dot{V}(\tilde{x}(t \mid t_{k+1})). \tag{2.40}$$

Integrating (2.40) from $t_k + T$ to $t_{k+1} + T$ yields:

$$\int_{t_k+T}^{t_{k+1}+T} \|\tilde{x}(t\,|t_{k+1})\|_{Q^*}^2 dt \le \|\tilde{x}(t_k + T\,|t_{k+1})\|_P^2 - \|\tilde{x}(t_{k+1} + T\,|t_{k+1})\|_P^2. \tag{2.41}$$

Substituting (2.41) into ΔV_1 yields:

$$\begin{aligned}\Delta V_1 &\le \|\tilde{x}(t_k + T\,|t_{k+1})\|_P^2 - \|\hat{x}^*(t_k + T\,|t_k)\|_P^2 \\ &\le (\|\tilde{x}(t_k + T\,|t_{k+1}) - \hat{x}^*(t_k + T\,|t_k)\|_P)(\|\tilde{x}(t_k + T\,|t_{k+1})\|_P \\ &\quad + \|\hat{x}^*(t_k + T\,|t_k)\|_P) \\ &\le \sigma_d e^{(1-\alpha)TG}(\varepsilon + \varepsilon_f),\end{aligned} \tag{2.42}$$

where, in the third inequality, (2.24) and the terminal constraint condition are applied. Since $\tilde{x}(t \mid t_{k+1})$ and $\hat{x}^*(t \mid t_k)$ satisfy the state constraints, define $b = \sup_{x\in\mathcal{X}}\|x\|_Q$ as the upper bound of the state norm. Then, it holds that:

$$\begin{aligned}\Delta V_2 \;&\le \int_{t_{k+1}}^{t_k+T} (\|\tilde{x}(t\,|t_{k+1} - \hat{x}^*(t\,|t_k\,\|_Q)(\|\tilde{x}(t\,|t_{k+1}\,\|_Q + \|\hat{x}^*(t\,|t_k\,\|_Q)\,dt \\ &\le \int_{t_{k+1}}^{t_k+T} 2b\sigma_d e^{G(t-t_{k+1})}\,dt \\ &\le \frac{2b\sigma_d}{G}(e^{(1-\alpha)TG} - 1).\end{aligned} \tag{2.43}$$

For ΔV_3, by applying the scaling method from [72], we obtain that:

$$\Delta V_3 \;\le -\frac{\underline{\lambda}(Q)}{\bar{\lambda}(P)} \int_{t_k}^{t_{k+1}} |\hat{x}^*(t|t_k)|_P^2 dt \le -\frac{\underline{\lambda}(Q)}{\bar{\lambda}(P)}\alpha T \varepsilon_f^2. \tag{2.44}$$

Substituting (2.42), (2.43), and (2.44) into (2.39), and combining with condition (2.38) yields:

$$J^*(x(t_{k+1})) - J^*(x(t_k)) \le -(1-\theta)\frac{\underline{\lambda}(Q)}{\bar{\lambda}(P)}\alpha T \varepsilon_f^2, \tag{2.45}$$

where

$$\theta = \frac{\sigma_d e^{(1-\alpha)TG}(\varepsilon + \varepsilon_f) + \frac{2b\sigma_d}{G}(e^{(1-\alpha)TG} - 1)}{\frac{\underline{\lambda}(Q)}{\bar{\lambda}(P)}\alpha T \varepsilon_f^2} \in (0, 1).$$

By applying recursion, we obtain:

$$J^*(x(t_k)) \leq J^*(x(t_0)) - k(1-\theta)\frac{\lambda(Q)}{\bar{\lambda}(P)}\alpha T\varepsilon_f^2. \tag{2.46}$$

Let

$$k^0 = \left\lceil \frac{J^*(x(t_0))}{(1-\theta)\frac{\lambda(Q)}{\bar{\lambda}(P)}\alpha T\varepsilon_f^2} \right\rceil.$$

Assume $x(t_{k^0+1}) \notin \Omega_\varepsilon$. Substituting $k = k^0 + 1$ into (2.46) yields $J^*(x(t_{k^0+1})) < 0$, which contradicts the fact that $J^*(x(t_{k^0+1})) \geq 0$. This contradiction shows that the assumption $x(t_{k^0+1}) \notin \Omega_\varepsilon$ is invalid. Therefore, the system state will converge to the set Ω_ε within finite time. Since each triggering interval does not exceed T, the upper bound of the finite convergence time can be estimated as $(k^0 + 1)T$.

If $x(t_k) \in \Omega_\varepsilon$, then by taking $V(x(t)) = \|x(t)\|_P^2$ as a Lyapunov candidate function and applying Lemma 2.1, we can obtain:

$$\dot{V}(x(t)) \leq -\|x(t)\|_{Q^*}^2 + 2x^T(t)Pw(t) \leq -\frac{\lambda(Q^*)}{\bar{\lambda}(P)}\|x(t)\|_P^2 + \frac{2\varepsilon\,\|P\|}{\sqrt{\underline{\lambda}(P)}}\|w(t)\|. \tag{2.47}$$

By applying the relevant results from Ref. [61], it is concluded that (2.47) indicates that the system is input-to-state stable (ISS) with respect to $w(t)$. □

In fact, according to Lemma 2.2, the actual triggering threshold will not exceed the desired triggering threshold σ_d (which is a function of α). Therefore, the feasibility and stability analysis are related to α. The parameter β is an adjustable periodic sampling value and is independent of the desired threshold σ_d. Hence, the choice of β does not affect feasibility and stability. However, the selection of β influences the triggering prediction value σ_2, and also affects the triggering interval, which will be verified in the simulation results.

2.4 Simulation Examples

2.4.1 Undamped Oscillator

Consider a nonlinear system [17]:

$$\dot{x}_1(t) = x_2(t) + u(t)(\mu - (1-\mu)x_1(t)) + w_1(t), \tag{2.48}$$

$$\dot{x}_2(t) = x_1(t) + u(t)(\mu - 4(1-\mu)x_2(t)) + w_2(t), \tag{2.49}$$

where $\mu = 0.99$, $x = [x_1, x_2]^T$ is the state vector, and $w = [w_1, w_2]^T \in \mathbb{R}^2$ is the bounded disturbance. The approximate linearization of this nonlinear system at the origin is stabilizable, thus satisfying Assumption 2.1. The control input constraint is $-2 \leq u(t) \leq 2$, and the system state constraints are $-1.2 \leq x_i(t) \leq 1.2$, $i = 1, 2$. The disturbance upper bound is known to be $\|w_i(t)\| \leq 0.005,\ i = 1, 2$; and in simulations, disturbances are generated randomly within this upper bound. The Lipschitz constant $G = 1.05$ is computed using the method from Ref. [73].

The state weighting matrix and control weighting matrix are set by $Q = \text{diag}\{1.2, 1.2\}$ and $R = 0.5$, respectively. Using the design method from Ref. [17], the state feedback gain is calculated by:

$$K = [-0.8, -0.8],$$

and the terminal weighting matrix is calculated by:

$$P = \begin{bmatrix} 5.31 & 4.11 \\ 4.11 & 5.31 \end{bmatrix}.$$

The parameters of the terminal set are chosen by $\varepsilon = 0.2$ and $\varepsilon_f = 0.15$ to satisfy Lemma 2.1 and Theorem 2.1. The prediction horizon is set to $T = 1.2$ s. The triggering threshold parameter α is set to 0.5, and the corresponding expected triggering threshold is computed by $\sigma_d = 6.4 \times 10^{-3}$. According to Theorems 2.1 and 2.2, the allowable disturbance upper bound η should be less than 7.1×10^{-3} to guarantee feasibility and stability, which is satisfied in this simulation example. The periodic sampling parameter is designed as $\beta = 0.25$. The initial state is suppose to be $x_0 = [0.8; 1]$.

The nonlinear system in this example is simulated using Algorithm 1. To compare the performance of different approaches, simulations are conducted using similar parameter settings for the two-phase event-triggered MPC proposed in this chapter, as well as for the event-triggered MPC in [51] and the self-triggered MPC in [61]. In event-triggered MPC, state information is continuously sampled, and the triggering condition is evaluated at each sampling instant. In contrast, self-triggered MPC determines the next triggering time based on measurements obtained at the current triggering instant, without requiring continuous sampling of the system state.

The state trajectories under different methods are shown in Figure 2.2. It can be observed that in all three strategies, the closed-loop system states converge to the neighborhood of the origin, and the convergence speeds are similar, with only slight differences. The corresponding control signals are plotted in Figure 2.3, which shows that all control constraints are satisfied, indicating the feasibility of the algorithm.

Due to the conservativeness introduced in the feasibility and stability analysis through scaling, the admissible disturbance bound is relatively small. In practice, however, the algorithm proposed in this chapter can maintain feasibility and stability under larger disturbances. Figure 2.4 illustrates the trajectories of the states

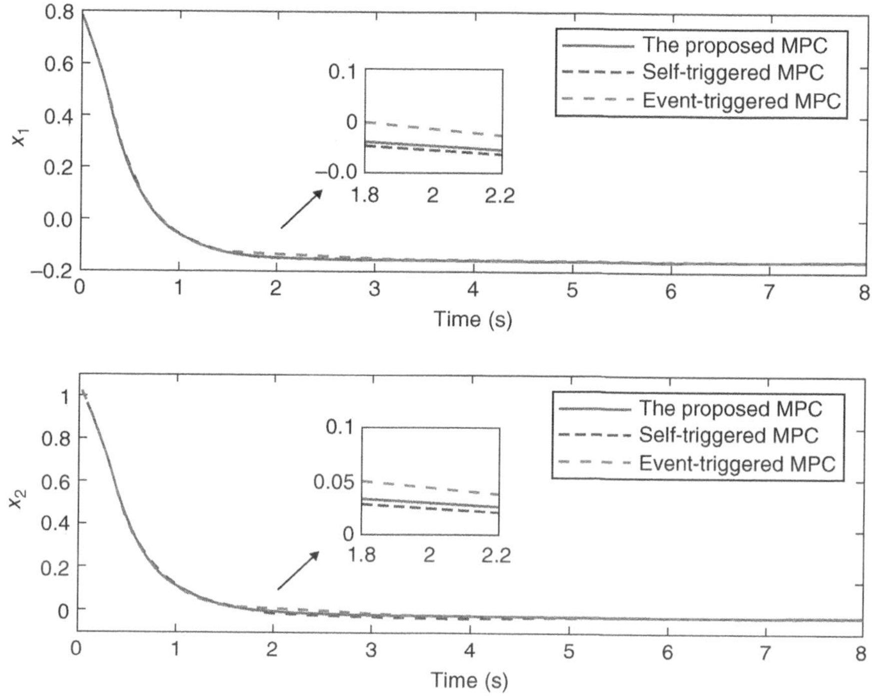

Figure 2.2 Trajectories of states under event-triggered MPC, self-triggered MPC, and two-phase event-triggered MPC.

and control inputs when the disturbance satisfies the condition $||w_i(t)|| \leq 0.12$. The results show that the system states still converge to the neighborhood of the origin under larger disturbances, and the control input gradually stabilizes and fluctuates around the origin.

To demonstrate the advantages of the proposed strategy, the sampling instants and computation instants are plotted in Figure 2.5, alongside comparisons with the corresponding instants for event-triggered MPC and self-triggered MPC. In Figure 2.5, solid markers represent computation instants, including optimization instants (denoted by dots) and local feedback instants (denoted by asterisks). Hollow markers indicate sampling-only instants where no optimization is performed.

At each sampling instant, the sensor measures the system state, and at each optimization instant, the processor solves a nonlinear optimization problem. Once the system enters the terminal region, a local state feedback controller is used, and no further optimization is processed.

Analysis of Figure 2.5 shows that the proposed two-phase event-triggered MPC strategy samples at times 0, 0.6, 0.9, 1.5, and 1.8 seconds then it switches to the

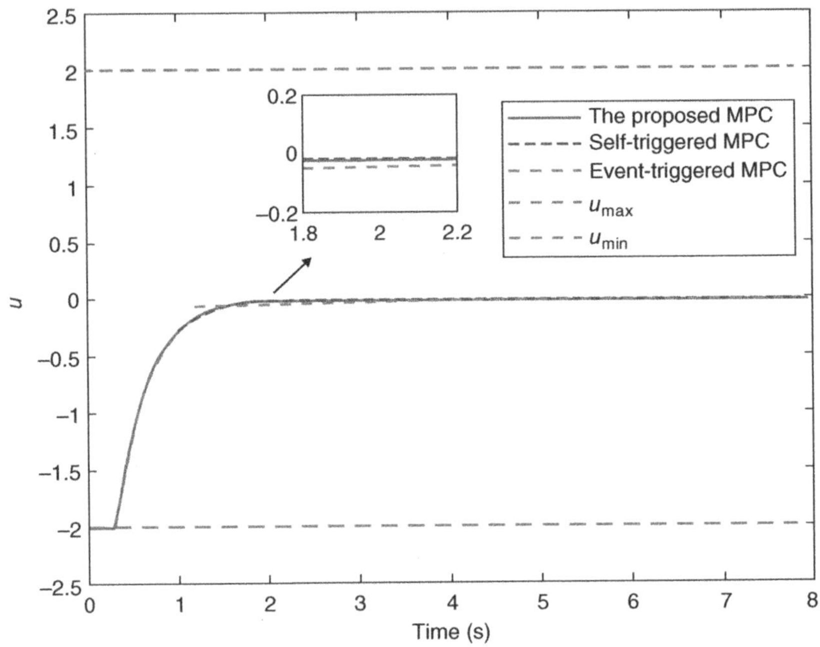

Figure 2.3 Trajectories of control inputs under event-triggered MPC, self-triggered MPC, and two-phase event-triggered MPC.

local state feedback controller. The optimization problem (2.4)–(2.8) is solved only at 0 and 0.9 seconds.

In comparison, the self-triggered MPC performs three trigger-and-optimize cycles, while the event-triggered MPC samples and checks the triggering condition 11 times. For all methods, the complexity of solving the optimization problem per trigger is approximately the same.

Due to the stochastic nature of disturbances, 30 simulations were conducted, and the average results were used for comparison and analysis. The results show that compared with the self-triggered MPC, the proposed strategy achieves lower triggering frequency. Compared with the event-triggered MPC, the proposed method requires fewer state measurements and less frequent evaluations of the triggering condition, which reduces sensor demands and power consumption.

Furthermore, if $\alpha = \beta = 0.1$, the proposed strategy degenerates into the periodic event-triggered MPC in [52]. If $\beta = 0$, it shares the same triggering instants as the self-triggered MPC in [61]. This demonstrates that the proposed approach is flexible and general, allowing adjustment of the measurement and triggering frequencies according to system performance and sensor characteristics.

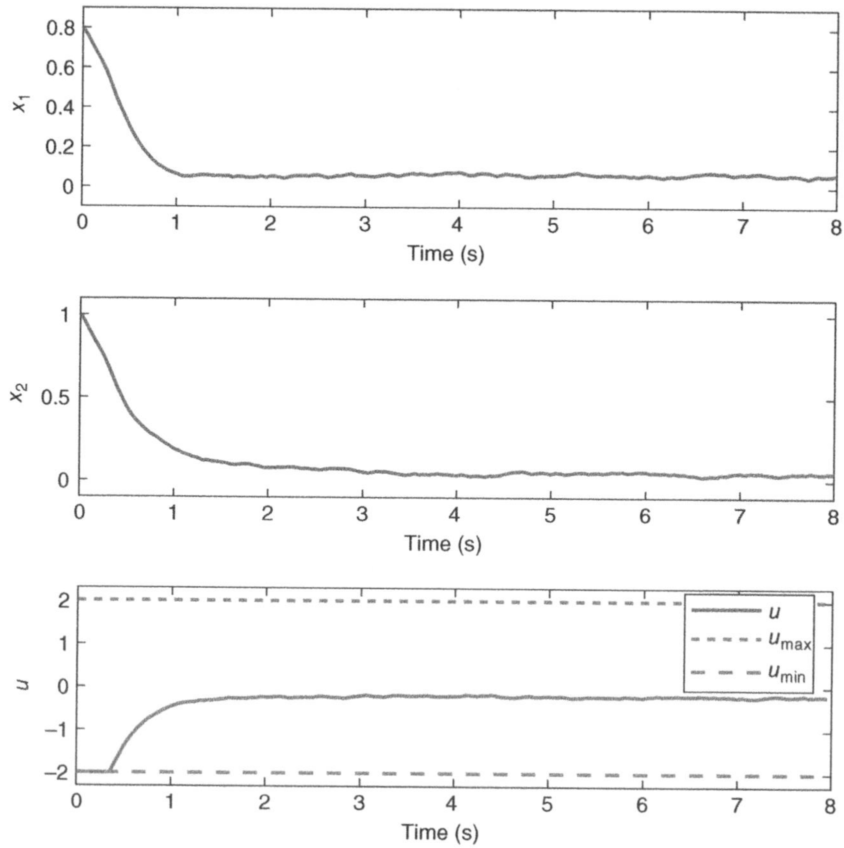

Figure 2.4 State and control input in case of large disturbances.

The desired triggering threshold σ_d can be adjusted by the parameter α. Table 2.1 compares the performance with different values of α, specifically $\alpha = 0.75$, $\alpha = 0.5$, and $\alpha = 0.25$. The optimal cost is defined by:

$$J^*_{\text{cost}}(x) = \int_0^8 \left(\|x(t)\|_Q^2 + \|u(t)\|_R^2 \right) dt.$$

From the analysis of Table 2.1, it can be observed that, as α increases, both the desired triggering threshold σ_d and the maximum allowable disturbance bound $\eta_{\max}$ increase accordingly. Moreover, a larger value of α can reduce the number of optimization problems solved, but it may also lead to a higher value of the optimal cost.

In Figure 2.5, when $\alpha = 0.5$, the proposed two-phase event-triggered MPC strategy solves the optimization problem twice. In contrast, the event-triggered MPC

Table 2.1 Performance with different values of α.

Values	σ_d	$\eta_{\max}$	Iterations of optimization	$J^*_{\text{cost}}(x)$
$\alpha = 0.75$	0.0202	0.0124	1	1.2072
$\alpha = 0.5$	0.0064	0.0071	2	1.2041
$\alpha = 0.25$	0.0019	0.0049	2	1.2029

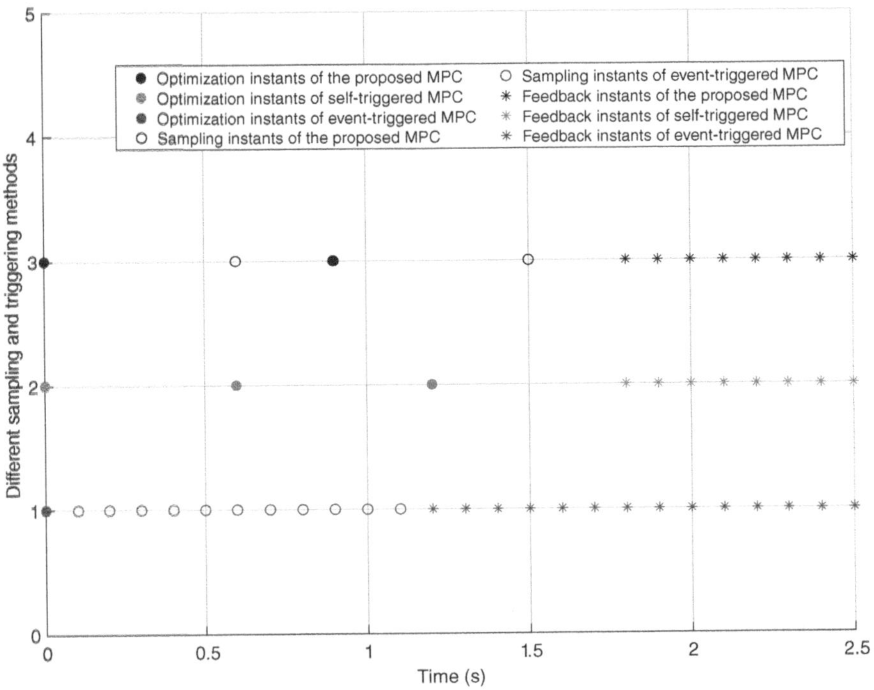

Figure 2.5 Sampling and optimization instants under event-triggered MPC, self-triggered MPC, and two-phase event-triggered MPC.

method solves it only once, which may suggest that the proposed strategy requires more computational resources. However, the number of times the optimization problem is solved can be adjusted by the parameter α. If $\alpha = 0.75$, the expected triggering interval increases, and the number of optimization problems solved decreases to once, as shown in Table 2.1. Under this condition, the computational burden of the proposed two-phase strategy is comparable to that of the event-triggered strategy. This demonstrates the flexibility of the proposed method, which allows for adjustment of computational load through parameter tuning.

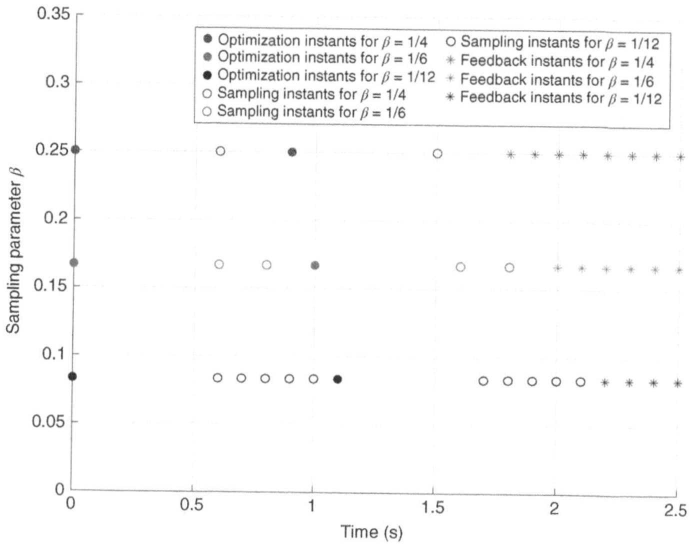

Figure 2.6 Sampling and optimization instants under different values of β.

The periodic sampling parameter β can be adjusted to achieve a balance between system performance and the frequency of measurement and decision-making. The sampling and triggering times for different values of β are illustrated in Figure 2.6. Analysis shows that as β decreases, the sampling frequency increases, thereby increasing the burden on sensors for measurements and on the processor for evaluating triggering conditions. Moreover, a smaller β results in longer intervals between optimization triggers, reducing the computational load within a fixed time period. As illustrated in Figure 2.6, when $\beta = 1/4$, the optimization problem is solved twice within 0.9 seconds, whereas for $\beta = 1/12$, it is solved twice within 1.1 seconds. Therefore, by appropriately tuning β, a balance can be achieved between computational load and the cost to sensors and the processor.

2.4.2 Simplified Spring-damper Element in Vehicle Suspension System

Considering a simplified spring-damper element in a vehicle suspension system:

$$\begin{cases} \dot{x}_1(t) = x_2(t), \\ \dot{x}_2(t) = -\dfrac{\xi_m}{M_m} e^{-x_1(t)} x_1(t) - \dfrac{h_d}{M_m} x_2(t) + \dfrac{u(t)}{M_m} + w(t), \end{cases} \tag{2.50}$$

where the position and velocity of the mass point are represented by $x_1(t)$ and $x_2(t)$, respectively. The mass of the suspension system block M_m is 1.25 kg, the spring stiffness ξ_m is 0.9 N/m, and the damping coefficient h_d is 0.42 Ns/m. The state constraints are $\|x_1(t)\| \leq 0.35$ m and $\|x_2(t)\| \leq 2$ m/s. It is supposed that the control input should satisfy $-0.9\ \text{N} \leq u(t) \leq 0.9$ N. The disturbance is bounded by $\|w(t)\| \leq 0.01\ \text{m/s}^2$, and it is generated randomly.

The two-phase event-based MPC is Algorithm 1 is applied to stabilize its origin. The Lipschitz constant is computed as $G = 1.46$. The weighting matrices are defined as $Q = \text{diag}\{0.6, 0.6\}$ and $R = 0.01$. The terminal weighting matrix is given by

$$P = \begin{bmatrix} 7.95 & 1.34 \\ 1.34 & 1.54 \end{bmatrix},$$

and the state feedback control law is calculated by $K = \begin{bmatrix} -4.67 & -6.45 \end{bmatrix}^T$. Additionally, the terminal parameters are set to $\varepsilon = 0.08$ and $\varepsilon_f = 0.072$. The prediction horizon is set to $T = 3$ s. The triggering threshold parameter α is defined as 0.4, and the periodic sampling parameter β is set to 0.1. The initial state is $x_1(0) = 0.3$ m, $x_2(0) = -0.8$ m/s.

With the strategy proposed in this chapter, the closed-loop states and control are illustrated by Figure 2.7. Analysis of this figure reveals that the system states converge toward the vicinity of the origin, while the control constraints are

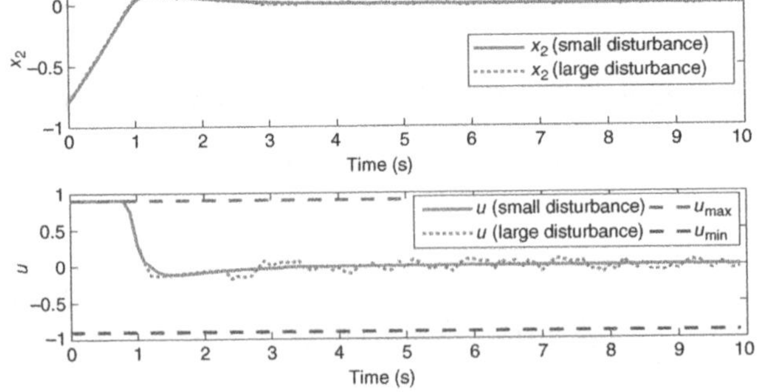

Figure 2.7 Closed-loop states and control in the presence of disturbances.

satisfied throughout the entire process. Due to the conservatism of the stability derivation, the actual system can maintain convergence properties even under larger disturbances. The system state and control trajectories under increased disturbances, where the disturbance satisfies $\|w(t)\| \leq 0.12\ \text{m/s}^2$, are plotted using orange dashed lines in Figure 2.7. It is observed that both the system states and control inputs gradually converge, exhibiting oscillations around the origin under the influence of the disturbance.

To highlight the advantages of the two-phase event-based MPC proposed in this chapter, the sampling and triggering instants, as well as those with other MPC strategies, are plotted in Figure 2.8. The self-triggered MPC estimates the next sampling instant based on the measurement obtained at the current triggering instant, and triggers directly at that time, eliminating the need for continuous state sampling. In contrast, the traditional time-triggered MPC samples the system state at fixed, small intervals and solves the optimization problem at each sampling instant. In Figure 2.8, solid dots represent the instants when the optimization problem is solved, asterisks denote local state feedback actions, and hollow dots indicate sampling without optimization computation. From Figure 2.8, it is evident that the method proposed in this chapter solves the optimization problem only once, the self-triggered MPC solves it twice, and the classical time-triggered MPC solves it 23 times.

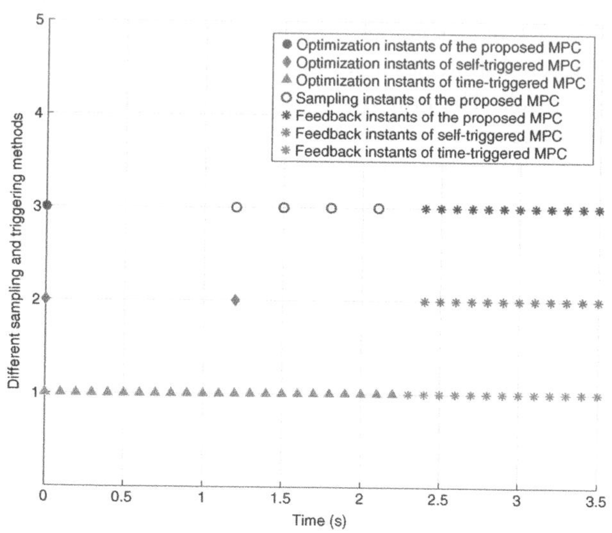

Figure 2.8 Sampling and triggering instants under different MPC methods.

Although the proposed strategy requires more sampling instants than the self-triggered method, due to the adjustability of sampling intervals, the additional sampling events are limited. Moreover, the computational resources required for these few extra sampling steps are negligible compared to those needed for solving optimization problems. Therefore, the method introduced in this chapter effectively reduces the number of optimizations and overall computational burden.

2.5 Conclusion

This chapter addresses the control problem of constrained continuous systems under external disturbances by proposing a two-phase event-triggered MPC strategy. In the first phase, the time interval required to reach a desired triggering threshold is calculated based on the worst-case impact of disturbances, denoted as αT. The optimal control signal obtained is applied to the actual system over this interval, during which no sampling is performed. After the first phase concludes, the second phase involves periodic sampling with an interval of βT. At each sampling instant, a triggering condition is evaluated, and a new trigger is initiated once the predicted safety threshold is exceeded. The proposed triggering strategy provides both upper and lower bounds for the triggering interval, thereby avoiding the occurrence of Zeno behavior. Compared with absolute event-triggered strategies, the method reduces the frequency of continuous sampling and condition evaluation. Compared with self-triggered strategies, it lessens estimation conservatism and extends the triggering intervals. The proposed mechanism can be seen as a generalized event-based MPC triggering strategy. Rigorous analyses of feasibility and stability are provided for the algorithm, and its effectiveness is demonstrated through simulation examples.

3

Event-triggered MPC for Discrete-time Systems with Aperiodic Sampling

In Chapter 2, a two-phase event-triggered model predictive control (MPC) strategy was proposed for continuous-time systems. Upper and lower bounds for the triggering interval were derived, ensuring that infinite triggering does not occur within finite time, thereby avoiding Zeno behavior. Compared to continuous-time systems, discrete-time systems inherently possess a minimum sampling interval, which preclude Zeno behavior. However, since their triggering intervals are discrete, a dedicated triggering mechanism must be designed.

In this chapter, for discrete-time disturbed nonlinear systems, an event-triggered aperiodic sampling MPC method is proposed. The algorithm designs a triggering threshold based on the convergence property of a Lyapunov function, and estimates the time interval required to reach that threshold by considering the worst-case effect of disturbances. After this interval elapses, sampling is performed, but triggering is not executed immediately. Instead, a new sampling interval is computed based on the deviation between the actual state at the sampling instant and the predicted state. This process is repeated until a sampling interval of zero is obtained, at which point the time is designated as a new triggering instant.

Classical self-triggered algorithms are often conservative because they rely on worst-case disturbance assumptions. The proposed algorithm shares a similar initialization step with self-triggered mechanisms but differs in that the computed interval is used as a sampling instant rather than a triggering instant. By leveraging feedback from the sampled state and the predicted state to recompute sampling intervals, the algorithm extends the triggering interval while still ensuring convergence.

This chapter presents further improvements to the aperiodic sampling strategy. First, the triggering condition is refined: instead of using a single state variable in the Lyapunov convergence inequality, it incorporates multiple, variable state components, thereby extending the triggering interval without significantly increasing computational cost. Second, a predictive horizon update strategy is introduced, incorporating a tunable contraction factor to alleviate the computational burden of solving the optimization problem at each step. Moreover, certain assumptions in the original algorithm are relaxed, including the removal of some Lipschitz continuity requirements. Finally, by constraining the disturbance bounds, it is proven that the system state will enter a robust invariant set.

The main contributions of this chapter include:

1. An event-triggered aperiodic sampling MPC is proposed for nonlinear systems with disturbances using a finite number of state samplings and evaluations. Compared to self-triggered MPC, it reduces conservativeness in the triggering estimation, extends triggering intervals, and lowers the frequency of solving optimization problems.
2. A predictive horizon update strategy is introduced with a tunable contraction factor to reduce computational load.
3. Rigorous proofs of recursive feasibility and closed-loop stability are provided for the proposed aperiodic sampling MPC algorithm.

3.1 Problem Statement

3.1.1 Plant to Be Controlled

Consider the discrete-time nonlinear system:

$$x(k+1) = f(x(k), u(k)) + w(k), \tag{3.1}$$

where $k \in \mathbb{N}$ denotes the time instant; $x(k) \in \mathbb{X} \subseteq R^n$ and $u(k) \in \mathbb{U} \subseteq R^m$ are the system state and control input, respectively; $\mathbb{X}$ and $\mathbb{U}$ are closed sets containing the origin; and $w(k) \in \mathbb{W} \subseteq \mathbb{R}^n$ is the external disturbance and satisfies $|w(k)| \leq \eta$.

Assumption 3.1 [74] The function $f(x, u)$ is Lipschitz continuous with respect to x, and satisfies $f(0, 0) = 0$. Let $h(x, u) = f(x, u) - x$, and denotes L_h as the Lipschitz constant of the function $h(x, u)$ with respect to x.

The corresponding undisturbed nominal system of (3.1) is given by:

$$\hat{x}(k+1) = f(\hat{x}(k), \hat{u}(k)). \tag{3.2}$$

The objective of this chapter is to design a control strategy that drives the system state to a neighborhood of the origin under disturbances, with a focus on satisfying constraints while improving computational efficiency. This is achieved through a dual-mode MPC framework, where an MPC law is applied outside the terminal region and a state feedback control law is used inside the terminal region.

3.1.2 Formulation of Optimization

Let the sequence k_l, $l \in \mathbb{N}$ represent the time instants where the optimization problem is solved. Then, at time k_l, the cost function to be optimized can be expressed by:

$$J(x(k_l), \hat{\mathbf{u}}(k_l)) = \sum_{i=0}^{N-1} L(\hat{x}(k_l + i|k_l), \hat{u}(k_l + i|k_l)) + V_f(\hat{x}(k_l + N|k_l)), \quad (3.3)$$

where N is the prediction horizon; $\hat{x}(k_l + i, |, k_l)$ and $\hat{u}(k_l + i, |, k_l)$ denote the predicted state and control input at time $k_l + i$ based on information at time k_l, respectively; and $\hat{\mathbf{u}}(k_l) = [\hat{u}(k_l, |, k_l), \hat{u}(k_l + 1, |, k_l), \dots, \hat{u}(k_l + N - 1, |, k_l)]$ represents the control input series. The stage cost is designed by:

$$L(\hat{x}(k_l + i, |, k_l), \hat{u}(k_l + i, |, k_l)) = |\hat{x}(k_l + i, |, k_l)|_Q^2 + |\hat{u}(k_l + i, |, k_l)|_R^2,$$

and the terminal cost is designed by:

$$V_f(\hat{x}(k_l + N, |, k_l)) = |\hat{x}(k_l + N|k_l)|_P^2,$$

where Q, R, and P are designed positive definite weighting matrices.

Assumption 3.2 [75] The stage cost function $L(x, u)$ is Lipschitz continuous with respect to $x \in \mathbb{X}$, and L_q is the corresponding Lipschitz constant.

Assumption 3.3 [61] There exists a set $\Omega_r = x \mid V_f(x) \leq r^2$ and a feedback control law $u = Kx$, such that $Kx \in \mathbb{U}$ for all $x \in \Omega_r$. Moreover, if the nominal system (3.2) is controlled using $u = K\hat{x}(k)$, then it satisfies:

$$V_f(\hat{x}(k + 1)) - V_f(\hat{x}(k)) \leq -|\hat{x}(k)|_{Q*}^2,$$

where $Q^* = Q + K^T RK$ is a positive definite matrix.

Under Assumption 3.3, the set Ω_r is a control invariant set for the nominal system under the control law $u = Kx$.

Based on the above definitions, the optimal control problem can be formulated. For the nominal system (3.2), the optimization problem can be constructed as follows:

$$\hat{\mathbf{u}}^*(k_l) = \arg \min_{\hat{u}(k_l+i|k_l)} J(x(k_l), \hat{\mathbf{u}}(k_l)), \quad (3.4)$$

$$\text{s.t. } \hat{x}(k_l \,|k_l) = x(k_l), \quad (3.5)$$

$$\hat{x}(k_l + i + 1 \,|k_l) = f(\hat{x}(k_l + i \,|k_l), \hat{u}(k_l + i \,|k_l)), \; 0 \leq i \leq N - 1, \quad (3.6)$$

$$\hat{x}(k_l + i \,|k_l) \in \mathbb{X} \ominus X_i, \quad 0 \leq i \leq N - 1, \quad (3.7)$$

$$\hat{u}(k_l + i \,|k_l) \in \mathbb{U}, \quad 0 \leq i \leq N - 1, \quad (3.8)$$

$$\hat{x}(k_l + N \,|k_l) \in \Omega_\varepsilon, \quad (3.9)$$

where

$$X_i = \left\{ x \,\middle|\, \|x\|_P \le \eta\sqrt{\bar{\lambda}(P)}\frac{(1+L_h)^i - 1}{L_h} \right\}$$

is the state contraction constraint set, accounting for possible effects of all disturbances. The nominal state set in (3.7) is obtained by subtracting the state contraction set X_i from the original state set $\mathbb{X}$. The set

$$\Omega_\varepsilon = \{x \,|V_f(x) \le \varepsilon^2, 0 < \varepsilon < r\}$$

is the terminal set. The optimal control sequence and the corresponding optimal state sequence obtained at time k_l are denoted by:

$$\hat{\mathbf{u}}^*(k_l) = [\hat{u}^*(k_l \,|k_l), \hat{u}^*(k_l + 1 \,|k_l), \dots, \hat{u}^*(k_l + N - 1 \,|k_l)],$$

$$\hat{\mathbf{x}}^*(k_l) = [\hat{x}^*(k_l \,|k_l), \hat{x}^*(k_l + 1 \,|k_l), \dots, \hat{x}^*(k_l + N \,|k_l)],$$

respectively. The obtained optimal control sequence is applied to the actual system (3.1) from time k_l until the next trigger time k_{l+1}. At time k_{l+1}, the optimization problem is resolved, and $\hat{x}(k_{l+1} \,|k_{l+1}) = x(k_{l+1})$ is assigned. When the system state enters the terminal set Ω_r, the controller is switched to local state feedback control, and the computation of the optimization problem is stopped.

Due to the presence of disturbances in actual nonlinear systems, a constraint-tightening method is adopted in the optimization problem design for the nominal system. In this method, a tightened constraint set for the state is constructed to account for the full effect of disturbances. The nominal state constraint set is obtained by subtracting the tightened set from the original state constraint set, and this nominal set is then used as the state constraint in the optimization problem. By applying the optimal control sequence obtained from this approach to the actual disturbed system, the resulting actual state will lie within the Minkowski sum of the nominal state set and the disturbance-induced deviation set, and therefore remain inside the original state constraint set.

Assumption 3.4 [61] The optimization problem (3.4)–(3.9) is solvable initially, and $\Omega_r \subseteq \mathbb{X} \ominus X_N$.

Due to the influence of bounded disturbances, there exists a certain deviation between the predicted state and the actual state. To analyze this deviation, Lemma 3.1 presents the discrete-time Gronwall inequality, and Lemma 3.2 derives an upper bound for the state deviation.

Lemma 3.1 *[76] Given that z_n, d_n, and g_n, $n \in \mathbb{N}$, are three nonnegative sequences, if $z_n \leq d_n + \sum_{0\leq k<n} g_k z_k$, then it holds that:*

$$z_n \leq d_n + \sum_{0\leq k<n} d_k g_k \prod_{k<j<n} (1 + g_j).$$

Furthermore, if $d_n = d$ is a constant, then

$$z_n \leq d \prod_{0\leq j<n} (1 + g_j).$$

Lemma 3.2 *Assume that the optimization problem (3.4)–(3.9) is solved at time step k_l, and the resulting optimal control sequence is applied to the actual disturbed system (3.1). Then, the deviation between the actual system state and the predicted state satisfies:*

$$\|x(k_l + j) - \hat{x}^*(k_l + j \mid k_l)\|_P \leq \phi(j), \quad j \leq N$$

where

$$\phi(j) = \frac{\eta\sqrt{\bar{\lambda}(P)}\left((1 + L_h)^j - 1\right)}{L_h}.$$

Proof: According to Assumption 3.1, the deviation between the actual state and the predicted state can be expressed as:

$$\begin{aligned}
&\|x\,(k_l + j) - \hat{x}^*\,(k_l + j|k_l)\,\|_P \\
&= \|x\,(k_l) + \sum_{i=0}^{j-1} (h\,(x\,(k_l + i), \hat{u}^*\,(k_l + i|k_l) + w\,(k_l + i)) \\
&-\hat{x}^*\,(k_l|k_l) - \sum_{i=0}^{j-1} h\,(\hat{x}^*\,(k_l + i|k_l), \hat{u}^*\,(k_l + i|k_l))\,\|_P \\
&\leq j\eta\sqrt{\bar{\lambda}\,(P)} + \sum_{i=0}^{j-1} L_h\|x\,(k_l + i) - \hat{x}^*\,(k_l + i|k_l)\,\|_P.
\end{aligned} \tag{3.10}$$

It then follows from Lemma 3.1 that:

$$\begin{aligned}
&\|x(k_l + j) - \hat{x}^*(k_l + j|k_l)\|_P \\
&\leq j\eta\sqrt{\bar{\lambda}(P)} + \eta\sqrt{\bar{\lambda}(P)}L_h \sum_{i=0}^{j-1} i(1 + L_h)^{j-1-i} \\
&\leq \eta\sqrt{\bar{\lambda}(P)}\frac{(1 + L_h)^j - 1}{L_h}.
\end{aligned} \tag{3.11}$$

According to the definition, we have $\phi(j) = \eta\sqrt{\bar{\lambda}(P)}\frac{(1+L_h)^j - 1}{L_h}$, therefore (3.11) satisfies:

$$\|x(k_l + j) - \hat{x}^*(k_l + j\,|k_l)\|_P \leq \phi(j), \quad j \leq N. \tag{3.12}$$

3.2 Aperiodic Triggering Mechanism

3.2.1 Triggering Mechanism

In this section, we propose an aperiodic dynamic triggering mechanism with state feedback. First, the triggering threshold and the minimum triggering interval are determined based on the system convergence properties. Since the minimum triggering interval is estimated under the worst-case disturbance scenario, the triggering threshold is usually not reached at this moment. Subsequently, an aperiodic sampling and detection strategy is adopted: at each sampling time, the state feedback information is used to inversely compute the time interval required to reach the triggering threshold. This interval is then used as the next sampling interval. The above process is repeated until the computed sampling interval becomes zero, in which case the last sampling time is taken as the new triggering instant.

Design the trigger threshold by $\phi(s_{k_l}) = \eta\sqrt{\bar{\lambda}(P)}\frac{(1+L_h)^{s_{k_l}} - 1}{L_h}$, where s_{k_l} is the minimum triggering interval. Based on the worst-case impact that disturbances might cause, and according to the desired convergence performance, the minimum triggering interval s_{k_l} can be calculated. Its expression is given by:

$$s_{k_l} = \min\{s^*_{k_l}, N\}, \tag{3.13}$$

$$s^*_{k_l} = \max\{z|\mu(z) \leq (1-\sigma)L(x(k_l), \hat{u}^*(k_l|k_l))\}, \tag{3.14}$$

where

$$\mu(z) = \eta\sqrt{\bar{\lambda}(P)}L_q \times \left(\frac{(1+L_h)^N - (1+L_h)^{N-z} - (1+L_h)^z + 1}{\sqrt{\underline{\lambda}(P)}L_h^2} + \frac{(1+L_h)^N - (1+L_h)^{N-z}}{L_h}(r+\varepsilon)\right);$$

$s^*_{k_l}$ is the triggering interval corresponding to the desired convergence performance, and its derivation is based on closed-loop stability (it will be specifically discussed in the stability section). The constant $\sigma \in (0, 1)$ is the convergence factor, which reflects the convergence performance of the system.

In order to ensure that the minimum triggering interval does not exceed the prediction horizon, the minimum value between $s^*_{k_l}$ and N in (3.13) should be selected as the minimum triggering interval s_{k_l}. Furthermore, to guarantee that the minimum triggering interval $s_{k_l} \geq 1$, it is necessary that $s^*_{k_l} \geq 1$ in (3.14). Substituting $z = 1$ into (3.14) and using inequality relations, a sufficient condition can be derived by:

$$\eta\sqrt{\bar{\lambda}(P)}L_q\left(\frac{(1+L_h)^{N-1}-1}{\sqrt{\underline{\lambda}(P)}L_h}+(1+L_h)^{N-1}(r+\varepsilon)\right)\leq(1-\sigma)\frac{\underline{\lambda}(Q)}{\bar{\lambda}(P)}r^2.$$

After a time interval of s_{k_l}, according to Lemma 3.2, the deviation between the actual state and the predicted state will reach $\phi(s_{k_l})$ under the worst-case impact of disturbances. Since this process considers the impact of the worst-case disturbance, there will typically be a certain distance between the actual state and the predicted state, and the desired triggering threshold after the time interval s_{k_l}. This distance can be used via an inverse mapping method to calculate a new sampling interval. At the new sampling instant, the state information is obtained, the deviation between the actual state and the predicted state is calculated, and a new sampling interval is determined through mapping. This process is repeated until the distance between the state deviation and the desired triggering threshold is zero or sufficiently small.

The triggering mechanism discussed in this chapter involves sampling the system state at the current moment, then calculating the next sampling moment. Between these two sampling moments, the optimal control sequence obtained by solving the optimization problem is applied to the actual system. This process is repeated until the calculated sampling interval becomes zero, at which point the last sampling moment serves as the next triggering moment. The calculation of the sampling interval is based on the inverse mapping principle, and it is described as follows.

At triggering instant k_l, the optimization problem (3.4)–(3.9) is solved to obtain the optimal control sequence. The sampling moment is initialized by $k_l^0 = k_l$, and the first sampling interval is $\gamma_l^0 = s_{k_l}$. The next sampling moment is then calculated by $k_l^1 = k_l^0 + \gamma_l^0$. Over the time interval $[k_l^0, k_l^1)$, the optimal control sequence is applied to the actual system, with no other sampling performed during this process.

Next, at time k_l^1, based on the state feedback information and using the deviation between the actual state $x(k_l^1)$ and the predicted state $\hat{x}^*(k_l^1 \mid k_l)$, the next sampling interval γ_l^1 corresponding to reaching the trigger threshold is calculated via an inverse mapping relationship:

$$\gamma_l^1 = s_{k_l} - \left\lceil \phi^{-1}\left(\left\| x(k_l^1) - \hat{x}^*(k_l^1 \mid k_l)\right\|_P\right)\right\rceil .$$

Then, the corresponding next sampling moment can be calculated by $k_l^2 = k_l^1 + \gamma_l^1$. The optimal control sequence is applied over the interval $[k_l^1, k_l^2)$ without additional sampling. At moment k_l^2, a new sampling is performed, and a new sampling interval is calculated. The aforementioned process is repeated: at moment k_l^i, a new sampling interval $\gamma_l^i = s_{k_l} - \lceil \phi^{-1}(\|x(k_l^i) - \hat{x}^*(k_l^i \mid k_l)\|_P) \rceil$ is calculated, along with the next sampling moment $k_l^{i+1} = k_l^i + \gamma_l^i$.

This sampling and calculation process is continuously repeated until, at moment k_l^n, γ_l^n is calculated to be zero. At that point, the next triggering instant can be expressed by $k_{l+1} = k_l^n$, and the interval between two triggering moments is denoted by $\tau_{k_l} = k_{l+1} - k_l$. The sampling and triggering process is illustrated by Figure 3.1.

If the calculated sampling interval $\gamma_l^n = 0$, it indicates that subsequent sampling would be unsafe and could potentially compromise the system convergence performance. Therefore, the final sampling time k_l^n is set as the new trigger time k_{l+1}. Since the prediction horizon is finite and state information beyond it is unknown, to ensure the algorithm's implementability, the maximum trigger interval cannot exceed the prediction horizon N.

Based on the analysis above, the triggering interval τ_{k_l} for the aperiodic sampling strategy can be expressed by:

$$\tau_{k_l} = \begin{cases} \sum_{i=0}^{n} \gamma_l^i, & if\ \gamma_l^n = 0, \sum_{i=0}^{n} \gamma_l^i < N, \\ N, & if\ \sum_{i=0}^{n} \gamma_l^i \geq N, \end{cases} \tag{3.15}$$

where

$$\gamma_l^i = s_{k_l} - \lceil \phi^{-1}(\|x(k_l^i) - \hat{x}^*(k_l^i \mid k_l)\|_P) \rceil$$

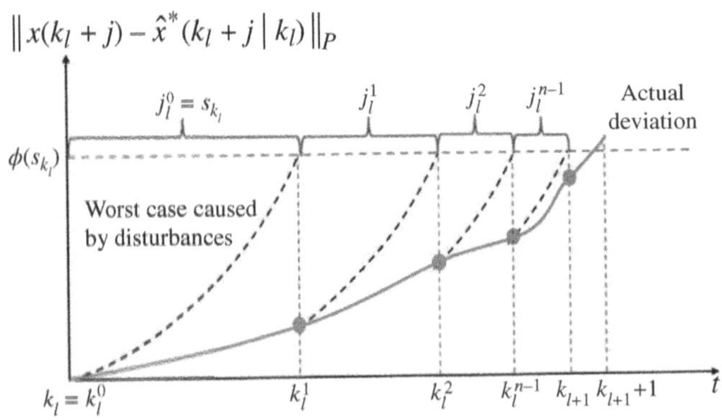

Figure 3.1 Illustration of aperiodic triggering mechanism.

is the sampling interval, and s_{k_l} is the minimum triggering interval. The sampling instants k_l^i satisfy $k_l^{i+1} = k_l^i + \gamma_l^i$ and $k_l^0 = k_l$.

The mechanism for selecting sampling and triggering instants is detailed in Algorithm 2. Once the system state enters the terminal set Ω_r, to reduce computational burden, the optimization problem is no longer solved, and triggering is ceased. Instead, a state feedback control law is employed for control. Therefore, the entire event-based aperiodic sampling MPC strategy can be comprehensively described using Algorithm 3.

Algorithm 2 Sampling and Triggering Instants

1: Initialize the sampling time $k_l^0 = k_l$. Calculate s_{k_l} via (3.13). Set the sampling interval $\gamma_l^0 = s_{k_l}$.

2: Calculate the next sampling time $k_l^1 = k_l^0 + \gamma_l^0$. Apply the optimal control sequence $\hat{\mathbf{u}}^*(k_l)$ to the actual system over the time interval $[k_l^0, k_l^1)$.

3: At time k_l^1, calculate the sampling interval $\gamma_l^1 = s_{k_l} - \lceil \phi^{-1}(\|x(k_l^1) - \hat{x}^*(k_l^1\,|k_l)\|_P)\rceil$.

- If $\gamma_l^1 = 0$, then the next trigger time is $k_{l+1} = k_l^1$, and the trigger interval is $\tau_{k_l} = k_l^1 - k_l$;
- If $\gamma_l^0 + \gamma_l^1 \geq N$, then the trigger interval is $\tau_{k_l} = N$;
- Otherwise, calculate the next sampling time $k_l^2 = k_l^1 + \gamma_l^1$. Apply the optimal control sequence over the interval $[k_l^1, k_l^2)$, set $i = 2$, and proceed to the next step.

4: Calculate the sampling interval $\gamma_l^i = s_{k_l} - \lceil \phi^{-1}(\|x(k_l^i) - \hat{x}^*(k_l^i\,|k_l)\|_P)\rceil$ and the sampling time $k_l^{i+1} = k_l^i + \gamma_l^i$.

- If $\gamma_l^i = 0$, then the next trigger time is set to $k_{l+1} = k_l^i$, and the trigger interval is expressed as $\tau_{k_l} = k_l^i - k_l$.
- If $\sum_{s=0}^{i} \gamma_l^s \geq N$, then the trigger interval is $\tau_{k_l} = N$.
- Otherwise, apply the optimal control sequence over the interval $[k_l^i, k_l^{i+1})$ and proceed to the next step.

5: Set $i = i + 1$. Repeat Step 4 until the sampling interval $\gamma_l^n = s_{k_l} - \lceil \phi^{-1}(\|x(k_l^n) - \hat{x}^*(k_l^n\,|k_l)\|_P)\rceil = 0$, where $n \in \mathbb{N}$, or until $\sum_{s=0}^{n} \gamma_l^n \geq N$.

- If $\gamma_l^n = 0$, set the next trigger moment as $k_{l+1} = k_l^n$, and simultaneously calculate the trigger interval $\tau_{k_l} = k_l^n - k_l$.
- If $\sum_{s=0}^{n} \gamma_l^n \geq N$, the trigger interval is set to $\tau_{k_l} = N$.

Algorithm 3 Event-based Aperiodic Sampling MPC Algorithm

1: At time instant k_l, measure the system state $x(k_l)$. If $x(k_l) \in \Omega_r$, apply the control law $u = Kx$. Otherwise, solve the optimization problem (3.4)–(3.9) to obtain the optimal control sequence $\hat{\mathbf{u}}^*(k_l)$ and the corresponding predicted state sequence $\hat{\mathbf{x}}^*(k_l)$.

2: Calculate the minimum trigger interval s_{k_l} via (3.13). Apply the optimal control sequence $\hat{\mathbf{u}}^*(k_l)$ to the actual system, ranging from $\hat{u}^*(k_l\,|k_l)$ up to $\hat{u}^*(k_l + s_{k_l} - 1\,|k_l)$ according to the corresponding time sequence. Let $k_l^0 = k_l$ and $\gamma_l^0 = s_{k_l}$, and calculate the sampling time $k_l^1 = k_l^0 + \gamma_l^0$. Set $i = 1$.

3: At time k_l^i, measure the system state $x(k_l^i)$, and calculate the next sampling interval γ_l^i using Algorithm 2.

4: If $\gamma_l^i \neq 0$ and $\sum_{s=0}^{i} \gamma_l^i \leq N$, apply the optimal control sequence $\hat{\mathbf{u}}^*(k_l)$ to the actual system, corresponding to time sequence from $\hat{u}^*(k_l^i\,|k_l)$ until $\hat{u}^*(k_l^i + \gamma_l^i - 1\,|k_l)$. Afterwards, update $k_l^{i+1} = k_l^i + \gamma_l^i$, set $i = i + 1$, and return to Step 3. If $\gamma_l^i = 0$ or $\sum_{s=0}^{i} \gamma_l^i \geq N$, update the next trigger moment to $k_{l+1} = k_l^i$, set $l = l + 1$, and return to Step 1.

If the triggering strategy proposed in this section is employed, the deviation between the actual state and the predicted state at the triggering time will not exceed the triggering threshold $\phi(s_{k_l})$. This conclusion is described by the following Lemma 3.3.

Lemma 3.3 *Suppose that k_l is the previous triggering time instant. If the aperiodic sampling triggering mechanism is applied, then the next triggering time satisfies:*

$$\|x(k_{l+1}) - \hat{x}^*(k_{l+1}\,|k_l)\|_P \leq \phi(s_{k_l}), \tag{3.16}$$

where $k_{l+1} = k_l + \tau_{k_l}$ is the next triggering time, and $\phi(s_{k_l})$ is the desired triggering threshold.

Proof: Given that k_l is the previous triggering time, and an event-based aperiodic sampling strategy is employed. According to the triggering algorithm, it holds that $k_l^1 = k_l^0 + \gamma_l^0 = k_l + s_{k_l}$. Directly applying the conclusion from Lemma 3.2 yields:

$$\left\|x(k_l^1) - \hat{x}^*(k_l^1\,|k_l)\right\|_P = \left\|x(k_l + s_{k_l}) - \hat{x}^*(k_l + s_{k_l}\,|k_l)\right\|_P \leq \phi(s_{k_l}). \tag{3.17}$$

Apply the mathematical induction here. Assume that at the sampling instant $k_l^i \in N_{[k_l,k_{l+1}]}$, the condition $\left\|x(k_l^i) - \hat{x}^*(k_l^i \mid k_l)\right\|_P \leq \phi(s_{k_l})$ is satisfied. Since:

$$\gamma_l^i = s_{k_l} - \left\lceil \phi^{-1}(\left\|x(k_l^i) - \hat{x}^*(k_l^i \mid k_l)\right\|_P)\right\rceil,$$

it holds that:

$$\left\|x(k_l^i) - \hat{x}^*(k_l^i \mid k_l)\right\|_P \leq \phi(s_{k_l} - \gamma_l^i). \tag{3.18}$$

At the next sampling instant k_l^{i+1}, according to (3.18) and Gronwall's inequality, we can obtain:

$$\begin{aligned} \|x(k_l^{i+1}) - \hat{x}^*(k_l^{i+1} \mid k_l)\|_P &\leq \phi(s_{k_l} - j_l^i) + j_l^i \eta \sqrt{\bar{\lambda}(P)} \\ &\quad + \sum_{s=0}^{j_l^i - 1} L_h \left\|x(k_l^i + s) - \hat{x}^*(k_l^i + s \mid k_l)\right\|_P \\ &\leq \phi(s_{k_l} - j_l^i) + \phi(j_l^i) \\ &\leq \phi(s_{k_l}). \end{aligned} \tag{3.19}$$

By induction, we can show that for $i \in \mathbb{N}_{[1,n]}$, the condition:

$$\left\|x(k_l^i) - \hat{x}^*(k_l^i \mid k_l)\right\|_P \leq \phi(s_{k_l})$$

holds. If $i = n$, according to the triggering mechanism, $k_{l+1} = k_l^n$, thus:

$$\|x(k_{l+1}) - \hat{x}^*(k_{l+1} \mid k_l)\|_P = \|x(k_l^n) - \hat{x}^*(k_l^n \mid k_l)\|_P \leq \phi(s_{k_l}). \tag{3.20}$$

which finishes the proof.

3.2.2 Stability and Feasibility

The feasibility analysis considers the existence and feasibility of solutions to the optimization problem (3.4)–(3.9). The fundamental idea is to prove, through mathematical induction, that if the optimization problem has a solution at the initial time, then a solution exists at all subsequent time instants.

Theorem 3.1 Consider the actual system (3.1), and suppose that Assumptions 3.1–3.4 hold. Apply the event-based aperiodic sampling MPC triggering strategy via Algorithms 2 and 3. If the disturbance bounds and terminal constraints satisfy the following conditions:

$$\eta \leq \frac{(r - \varepsilon)L_h}{\sqrt{\bar{\lambda}(P)}(1 + L_h)^N}, \quad \left(1 - \frac{\underline{\lambda}(Q^*)}{\bar{\lambda}(P)}\right) r^2 \leq \varepsilon^2, \tag{3.21}$$

then the optimization problem (3.4–3.9) is recursively feasible.

Proof: According to Assumption 3.4, the optimization problem (3.4)–(3.9) has a solution at the initial time k_0. Assume that the optimization problem has a solution at time k_l, where $l \geq 0$. To prove feasibility, we need to demonstrate that at the next time the optimization problem is solvable, i.e., at the triggering time k_{l+1}, there exists a feasible control sequence such that the terminal constraint, state constraints, and control constraints are all satisfied.

Consider the case where the trigger interval $\tau_{k_l} < N$. Utilizing the optimal control sequence solved at time k_l and the terminal state feedback control law, a feasible control sequence for time k_{l+1} can be constructed by:

$$\tilde{u}(k_{l+1}+i\,|k_{l+1}) = \begin{cases} \hat{u}^*(k_{l+1}+i\,|k_l), & i \in [0, N-\tau_{k_l}-1], \\ K\tilde{x}(k_{l+1}+i\,|k_{l+1}), & i \in [N-\tau_{k_l}, N). \end{cases} \tag{3.22}$$

The corresponding state sequence is represented by $\tilde{x}(k_{l+1}+i\,|k_{l+1})$, where $\tilde{x}(k_{l+1}\,|k_{l+1}) = x(k_{l+1})$.

First, we analyze the terminal constraint. Consider the interval $[k_{l+1}, k_l + N]$. For $i \in \mathbb{N}_{[0,N-\tau_{k_l}]}$, the deviation between the feasible state and the optimal state can be expressed by:

$$\begin{aligned} \|\tilde{x}(k_{l+1}+i\,|k_{l+1}) - \hat{x}^*(k_{l+1}+i\,|k_l)\|_P &\leq \|\tilde{x}(k_{l+1}\,|k_{l+1}) - \hat{x}^*(k_{l+1}\,|k_l)\|_P \\ &\quad + \sum_{s=0}^{i-1} L_h \|\tilde{x}(k_{l+1}+s\,|k_{l+1}) - \hat{x}^*(k_{l+1}+s\,|k_l)\|_P. \end{aligned} \tag{3.23}$$

From (3.16) of Lemma 3.3, it holds that $\|\tilde{x}(k_{l+1}\,|k_{l+1}) - \hat{x}^*(k_{l+1}\,|k_l)\|_P \leq \phi(s_{k_l})$. It then follows from (3.23) and Lemma 3.1 that:

$$\|\tilde{x}(k_{l+1}+i\,|k_{l+1}) - \hat{x}^*(k_{l+1}+i\,|k_l)\|_P \leq \eta\sqrt{\bar{\lambda}(P)}\frac{(1+L_h)^{s_{k_l}}-1}{L_h}(1+L_h)^i. \tag{3.24}$$

Substitute $i = N - \tau_{k_l}$ into (3.24), and consider that $N > \tau_{k_l} \geq s_{k_l}$. It follows from (3.9) and (3.21) that:

$$\|\tilde{x}(k_l+N\,|k_{l+1})\|_P \leq \varepsilon + r - \varepsilon \leq r, \tag{3.25}$$

indicating that $\tilde{x}(k_l+N\,|k_{l+1}) \in \Omega_r$.

Consider the interval $[k_l + N, k_{l+1} + N]$. According to Assumption 3.3, for $i \in \mathbb{N}_{[0,\tau_{k_l}-1]}$, the following condition is satisfied:

$$\begin{aligned} &V_f(\tilde{x}(k_l + N + i + 1\,|k_{l+1})) - V_f(\tilde{x}(k_l + N + i\,|k_{l+1})) \\ &\leq -\|\tilde{x}(k_l + N + i\,|k_{l+1})\|_{Q^*}^2 \\ &\leq -\frac{\underline{\lambda}(Q^*)}{\bar{\lambda}(P)} V_f(\tilde{x}(k_l + N + i\,|k_{l+1})). \end{aligned} \tag{3.26}$$

Rearranging (3.26) yields the relationship between $V_f(\tilde{x}(k_l + N + i + 1\,|k_{l+1}))$ and $V_f(\tilde{x}(k_l + N + i\,|k_{l+1}))$ as:

$$V_f(\tilde{x}(k_l + N + i + 1\,|k_{l+1})) \leq \left(1 - \frac{\underline{\lambda}(Q^*)}{\bar{\lambda}(P)}\right) V_f(\tilde{x}(k_l + N + i\,|k_{l+1})). \tag{3.27}$$

Multiplying (3.27) from $i = 0$ to $i = \tau_{k_l} - 1$ yields:

$$V_f(\tilde{x}(k_l + N + \tau_{k_l}\,|k_{l+1})) \leq \left(1 - \frac{\underline{\lambda}(Q^*)}{\bar{\lambda}(P)}\right)^{\tau_{k_l}} V_f(\tilde{x}(k_l + N\,|k_{l+1})). \tag{3.28}$$

It follows from $\tau_{k_l} \geq 1$ and (3.25) that $\|\tilde{x}(k_l + N\,|k_{l+1})\|_P \leq r$. Substituting $k_{l+1} = k_l + \tau_{k_l}$ into (3.28) and utilizing condition (3.21) yields:

$$V_f(\tilde{x}(k_{l+1} + N\,|k_{l+1})) \leq \left(1 - \frac{\underline{\lambda}(Q^*)}{\bar{\lambda}(P)}\right) r^2 \leq \varepsilon^2, \tag{3.29}$$

implying that $\tilde{x}(k_{l+1} + N\,|k_{l+1}) \in \Omega_\varepsilon$, and the terminal constraint is satisfied.

Consider the state constraints. For $i \in N_{[0,N-\tau_{k_l}]}$, it follows from (3.7) that $\hat{x}^*(k_{l+1} + i\,|k_l) \in \mathbb{X} \ominus X_{i+\tau_{k_l}}$. Combining this with (3.24) yields:

$$\tilde{x}(k_{l+1} + i\,|k_{l+1}) \in \mathbb{X} \ominus X_{i+\tau_{k_l}} \oplus B\eta\sqrt{\bar{\lambda}(P)}\frac{(1 + L_h)^{s_{k_l}} - 1}{L_h}(1 + L_h)^i, \tag{3.30}$$

where $B(\theta) \triangleq \{x : \|x\|_P \leq \theta\}$.

It follows from $\tau_{k_l} \geq s_{k_l}$ that:

$$X_i \subseteq X_{i+\tau_{k_l}} \ominus B\eta\sqrt{\bar{\lambda}(P)}\frac{(1 + L_h)^{s_{k_l}} - 1}{L_h}(1 + L_h)^i,$$

It then follows from (3.30) that:

$$\tilde{x}(k_{l+1} + i\,|k_{l+1}) \in \mathbb{X} \ominus X_i.$$

Consequently, for $i \in N_{[0,N-\tau_{k_l}]}$, state constraints are satisfied.

For $i \in \mathbb{N}_{[N-\tau_{k_l}+1,N]}$, by Assumption 3.3, it is known that the predicted state using the state feedback control law

$$\tilde{u}(k_{l+1}+i\,|k_{l+1}) = K\tilde{x}(k_{l+1}+i\,|k_{l+1})$$

satisfies $\tilde{x}(k_{l+1}+i+1\,|k_{l+1}) \in \Omega_r$. Furthermore, according to Assumption 3.4, $\Omega_r \subseteq X \ominus X_N$, therefore for $i \in \mathbb{N}_{[N-\tau_{k_l}+1,N]}$ it holds that:

$$\tilde{x}(k_{l+1}+i\,|k_{l+1}) \in \mathbb{X} \ominus X_N \subseteq \mathbb{X} \ominus X_i.$$

In summary, for $i \in \mathbb{N}_{[0,N]}$, the state constraints are satisfied.

Based on the construction process of the feasible control sequence (3.22) and Assumption 3.3, it can be seen that the control constraints are satisfied on the interval $[k_l+N, k_{l+1}+N]$.

If the triggering interval satisfies $\tau_{k_l} = N$, then the feasible control sequence can be constructed as $\tilde{u}(k_{l+1}+i\,|k_{l+1}) = K\tilde{x}(k_{l+1}+i\,|k_{l+1})$, $i \in [0, N)$. This control sequence can be regarded as a special case of (3.22). Using a similar analysis method, it can be proven that all constraint conditions are satisfied.

Based on the above analysis, the feasible control sequence constructed at time k_{l+1} satisfies the terminal constraints, state constraints, and control constraints. Therefore, the optimization problem is feasible recursively. □

Closed-loop stability is given by the following theorem.

Theorem 3.2 Suppose that the system satisfies Assumptions 3.1–3.4 and the conditions in Theorem 3.1. With the event-triggered aperiodic sampling model predictive control Algorithms 2 and 3, the system state converges to the control invariant set Ω_r within a finite time. Furthermore, once the state enters the set Ω_r, the system is input-to-state stable with respect to disturbances.

Proof: Suppose that, at the initial time k_0, the system state is outside Ω_r. Define

$$V(x(k_l)) = J(x(k_l), \hat{\mathbf{u}}^*(k_l)) = \min_{\hat{u}(k_l+i|k_l)} J(x(k_l), \hat{\mathbf{u}}(k_l))$$

as the candidate Lyapunov function when using the optimal control sequence $\hat{\mathbf{u}}^*(k_l)$ at time $x(k_l)$, $l \geq 0$. The cost function corresponding to the feasible state sequence at time $x(k_{l+1})$ is defined by:

$$\tilde{V}(x(k_{l+1})) = J(x(k_{l+1}), \tilde{u}(k_{l+1})).$$

Based on the optimality of the cost function, it holds that:

$$V(x(k_{l+1})) \leq \tilde{V}(x(k_{l+1})).$$

Then the deviation between $V(x(k_{l+1}))$ and $V(x(k_l))$ can be expressed by:

$$\begin{aligned} V(x(k_{l+1})) - V(x(k_l)) &\leq J(x(k_{l+1}), \tilde{u}(k_{l+1})) - J(x(k_l), \hat{\mathbf{u}}^*(k_l)) \\ &\leq \Delta V_1 + \Delta V_2 + \Delta V_3, \end{aligned} \tag{3.31}$$

$$\begin{aligned} \Delta V_1 &= -\sum_{i=0}^{\tau_{k_l}-1} (\|\hat{x}^*(k_l+i\,|k_l)\|_Q^2 + \|\hat{u}^*(k_l+i\,|k_l)\|_R^2), \\ \Delta V_2 &= \sum_{i=\tau_{k_l}}^{N-1} (\|\tilde{x}(k_l+i\,|k_{l+1})\|_Q^2 - \|\hat{x}^*(k_l+i\,|k_l)\|_Q^2), \\ \Delta V_3 &= \sum_{i=N}^{N+\tau_{k_l}-1} (\|\tilde{x}(k_l+i\,|k_{l+1})\|_Q^2 + \|\tilde{u}(k_l+i\,|k_{l+1})\|_R^2) \\ &\quad + \|\tilde{x}(k_{l+1}+N\,|k_{l+1})\|_P^2 - \|\hat{x}^*(k_l+N\,|k_l)\|_P^2. \end{aligned}$$

For ΔV_1, it is calculated by:

$$\begin{aligned} \Delta V_1 &= -\sum_{i=0}^{\tau_{k_l}-1} (\|\hat{x}^*(k_l+i\,|k_l)\|_Q^2 + \|\hat{u}^*(k_l+i\,|k_l)\|_R^2) \\ &\leq -L(x(k_l), \hat{u}^*(k_l\,|k_l)). \end{aligned} \tag{3.32}$$

For ΔV_2, it follows from $\tau_{k_l} \geq s_{k_l}$, Assumption 3.2 and (3.24) that:

$$\begin{aligned} \Delta V_2 &\leq \sum_{i=\tau_{k_l}}^{N-1} L_q \|\tilde{x}(k_l+i\,|k_{l+1}) - \hat{x}^*(k_l+i\,|k_l)\| \\ &\leq L_q \eta \sqrt{\bar{\lambda}(P)} \frac{(1+L_h)^N - (1+L_h)^{N-s_{k_l}} - (1+L_h)^{s_{k_l}} + 1}{\sqrt{\underline{\lambda}(P)} L_h^2}. \end{aligned} \tag{3.33}$$

It then follows from Assumption 3.3 that:

$$\begin{aligned} &\|\tilde{x}(k_l+N+i+1\,|k_{l+1})\|_P^2 - \|\tilde{x}(k_l+N+i\,|k_{l+1})\|_P^2 \\ &\leq -\|\tilde{x}(k_l+N+i\,|k_{l+1})\|_Q^2 - \|\tilde{u}(k_l+N+i\,|k_{l+1})\|_R^2. \end{aligned} \tag{3.34}$$

Summation of (3.34) from $i = 0$ to $i = \tau_{k_l} - 1$ yields:

$$\begin{aligned} &\|\tilde{x}(k_{l+1}+N\,|k_{l+1})\|_P^2 - \|\tilde{x}(k_l+N\,|k_{l+1})\|_P^2 \\ &\leq -\sum_{i=N}^{N+\tau_{k_l}-1} (\|\tilde{x}(k_l+i\,|k_{l+1})\|_Q^2 + \|\tilde{u}(k_l+i\,|k_{l+1})\|_R^2). \end{aligned} \tag{3.35}$$

For ΔV_3, substituting (3.24) and (3.25) and the terminal constraint (3.9) into (3.35) yields:

$$\Delta V_3 \leq \|\tilde{x}(k_l + N \,|k_{l+1})\|_P^2 - \|\hat{x}^*(k_l + N \,|k_l)\|_P^2$$
$$\leq \eta\sqrt{\bar{\lambda}(P)}\frac{(1+L_h)^N - (1+L_h)^{N-s_{k_l}}}{L_h}(r+\varepsilon). \tag{3.36}$$

Substituting (3.32), (3.33), and (3.36) into (3.31), and combining the minimum triggering condition (3.13) and (3.14) yields:

$$V(x(k_{l+1})) - V(x(k_l)) \leq -\sigma L(x(k_l), \hat{u}^*(k_l \,|k_l)) \leq -\sigma\frac{\underline{\lambda}(Q)}{\bar{\lambda}(P)}\varepsilon^2. \tag{3.37}$$

Recursive summation of (3.37) yields:

$$V(x(k_{l+1})) \leq V(x(k_0)) - (l+1)\sigma\frac{\underline{\lambda}(Q)}{\bar{\lambda}(P)}\varepsilon^2. \tag{3.38}$$

Suppose (by contradiction) that the system state cannot converge to the terminal set Ω_r within a finite time. Then, from (3.38), as l approaches infinity, $V(x(k_{l+1}))$ would be less than zero. However, by definition, $V(x(k_{l+1})) \geq 0$, which leads to a contradiction. Therefore, the system state converges to the terminal set Ω_r within a finite time.

Once the system state enters the set Ω_r, the state feedback control law $u(k) = Kx(k)$ is applied to the real system. Using $V_f(x(k)) = \|x(k)\|_P^2$ as a candidate Lyapunov function and considering Assumption 3.3, we can obtain:

$$\begin{aligned}
&V_f(x(k+1)) - V_f(x(k)) \\
&= \|f(x(k), Kx(k)) + w(k)\|_P^2 - \|x(k)\|_P^2 \\
&\leq -\|x(k)\|_{Q^*}^2 + 2w^T(k)Pf(x(k), Kx(k)) + \|w(k)\|_P^2 \\
&\leq -\frac{\underline{\lambda}(Q^*)}{\bar{\lambda}(P)}\|x(k)\|_P^2 + \frac{2\eta r\,\|P\|}{\sqrt{\underline{\lambda}(P)}} + \bar{\lambda}(P)\eta^2.
\end{aligned} \tag{3.39}$$

It then follows from the process in [61] that (3.39) indicates that the closed-loop system is input-to-state stable with respect to bounded disturbances. □

Remark 3.1 Once the system state enters the terminal set Ω_r, if the upper bound of the disturbance satisfies certain conditions, it can be further proved that this set is a robust invariant set. Related conclusions will be specifically discussed in a later section.

3.3 Improved Aperiodic Triggering Mechanism

This section further improves upon the aperiodic triggering mechanism proposed in Section 3.2. Firstly, for the optimization problem, a predictive step size update strategy is adopted, and by introducing a time-domain contraction factor, the computational burden of solving a single optimization problem is reduced. Second, the triggering condition is improved; without significantly increasing the computational load, the single-state variable in the triggering interval condition is changed to multiple variable state variable information, thereby increasing the minimum trigger interval while ensuring convergence. Additionally, the conditions of the previous algorithms are relaxed by removing the Lipschitz condition in Assumption (3.2). Finally, when analyzing system stability, by limiting the perturbation boundary, it is proven that the final state will enter the robust invariant set, thus yielding a stronger stability conclusion.

3.3.1 Statement of Improved Triggering Mechanism

The objective of this section is to refine the existing triggering mechanism methodology for the perturbed system (3.1) and, through a MPC approach, ensure the convergence of the system state to the vicinity of the origin.

The first refinement involves adopting a prediction horizon update strategy, which, through the introduction of a time-domain contraction factor, reduces the prediction horizon of the optimization problem, thereby alleviating the computational burden of solving each optimization problem. Let $\{k_l\}, l \in \mathbb{N}$ denotes the sequence of optimization problem solution instants, and N_{k_l} represents the prediction horizon for each optimization problem. The cost function to be optimized is reformulated by:

$$J(x(k_l), \hat{\mathbf{u}}(k_l)) = \sum_{i=0}^{N_{k_l}-1} L(\hat{x}(k_l + i \,|k_l), \hat{u}(k_l + i \,|k_l)) + V_f(\hat{x}(k_l + N_{k_l} \,|k_l)). \tag{3.40}$$

Compared to the cost function in (3.3), the prediction horizon N_{k_l} in (3.40) is variable. The corresponding optimization problem is reformulated by:

$$\hat{\mathbf{u}}^*(k_l) = \arg \min_{\hat{u}(k_l+i|k_l)} J(x(k_l), \hat{\mathbf{u}}(k_l)), \tag{3.41}$$

$$\text{s.t.} \quad \hat{x}(k_l \,|k_l) = x(k_l), \tag{3.42}$$

$$\hat{x}(k_l + i + 1 \,|k_l) = f(\hat{x}(k_l + i \,|k_l), \hat{u}(k_l + i \,|k_l)), \; 0 \le i \le N_{k_l} - 1, \tag{3.43}$$

$$\hat{u}(k_l + i \,|k_l) \in \mathbb{U}, \quad 0 \le i \le N_{k_l} - 1, \tag{3.44}$$

$$\hat{x}(k_l + i\,|k_l) \in \mathbb{X} \ominus X_i, \quad 0 \leq i \leq N_{k_l} - 1, \tag{3.45}$$

$$\hat{x}(k_l + N_{k_l}\,|k_l) \in \Omega_\varepsilon. \tag{3.46}$$

Compared to the optimization problem in (3.4)–(3.9), the prediction horizon N_{k_l} in the new optimization problem (3.41)–(3.46) is variable, thus altering the computational burden of solving each instance of the optimization problem. If the prediction horizon exhibits a decreasing trend, the computational load for each optimization problem solution will progressively diminish, thereby reducing the computational burden. A crucial aspect is determining how to reasonably update the prediction horizon to ensure the feasibility and solvability of the optimization problem, and consequently, the stability of the closed-loop system.

For the prediction horizon update strategy, there are primarily two strategies. The first involves incorporating the prediction horizon as a decision variable into the optimization problem, thereby formulating a mixed-integer programming problem. Solving this programming problem yields the optimal prediction horizon [77]. Although this method can determine the prediction horizon corresponding to optimal performance, solving such an optimization problem is highly complex, consumes significant computational resources, and leads to an excessive computational burden. The alternative strategy utilizes the characteristics of predictive control to design the update strategy [78]. Inspired by Ref. [78], this section proposes a novel prediction horizon update strategy. The core of this strategy is that as the state gradually approaches the terminal region, the prediction horizon can be adjusted based on the triggering interval τ_{k_l}, thereby achieving the objective of reducing the prediction horizon. The prediction horizon update strategy is detailed below.

Let k_l and k_{l+1} denote two successive triggering instants, and define the triggering interval by $\tau_{k_l} = k_{l+1} - k_l$. The rationale behind the prediction horizon updating strategy is as follows. Under the effect of the optimal control sequence obtained at the previous triggering instant k_l, the system state gradually approaches the terminal set. This behavior implies that the state has progressed a certain distance along the desired trajectory. Accordingly, at the subsequent triggering instant k_{l+1}, the remaining time required for the system state to reach the terminal set can be reduced.

From the perspective of the optimization problem, this means that at the new triggering instant, the number of prediction steps required to solve the optimization problem can be decreased without degrading the control performance. To this end, a temporal shrinking factor $\alpha \in (0, 1)$ is introduced, and $\alpha\tau_{k_l}$ is employed to determine the amount that the prediction horizon can be reduced. Moreover, to guarantee both the control performance and the nonvanishing prediction length, one additional prediction step is included.

In summary, the prediction horizon updating strategy can be formulated by:

$$N_{k_{l+1}} = \min\{N_{k_l}, N_{k_l} - \lfloor \alpha\tau_{k_l} \rfloor + 1\}. \tag{3.47}$$

To provide an intuitive illustration of the prediction horizon updating strategy, the relationship between successive prediction horizons is depicted in Figure 3.2.

The second improvement of the triggering mechanism lies in the refinement of the triggering condition. In the triggering condition and subsequent analysis presented in the previous section, the Lipschitz continuity assumption on the stage cost function, as stated in Assumption 3.2, was employed. However, the Lipschitz constant L_q is defined over the entire state space, which introduces a certain degree of conservatism. In contrast, the improved triggering mechanism discussed in this section relaxes the requirement and no longer relies on the Lipschitz condition in Assumption 3.2.

Furthermore, provided that convergence performance is guaranteed, the triggering mechanism is enhanced by replacing the single-state-based triggering rule with an aggregated rule involving multistate information. This modification enlarges the minimum inter-event interval without significantly increasing the computational burden.

Considering the worst-case effect of disturbances and aiming to ensure convergence properties, the minimum inter-event interval $s_{k_l}^{\text{new}}$ can be constructed by:

$$\begin{aligned} s_{k_l}^{\text{new}} = \sup \Bigg\{ z \in \mathbb{N} : & \sum_{i=z}^{N_{k_l}-1} (\beta^2(i) + 2\beta(i) \, \|\hat{x}^*(k_l + i \,|k_l\|_Q) \\ & + \eta\sqrt{\bar{\lambda}(P)} \frac{(1+L_h)^{N_{k_l}} - (1+L_h)^{N_{k_l}-z}}{L_h} (r + \varepsilon) \\ & \leq (1-\sigma) \sum_{i=0}^{z-1} (\|\hat{x}^*(k_l + i \,|k_l\|_Q^2 + \|\hat{u}^*(k_l + i \,|k_l)\|_R^2) \Bigg\}, \end{aligned} \tag{3.48}$$

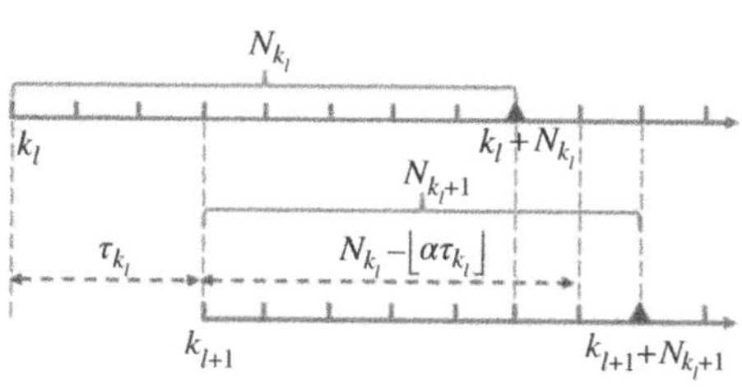

Figure 3.2 Prediction horizons of successive triggering instants.

where $\beta(i) \triangleq \eta\sqrt{\bar{\lambda}(Q)}\frac{(1+L_h)^z - 1}{L_h(1+L_h)^z}(1+L_h)^i$ with $\sigma \in (0,1)$ reflects the desired convergence performance.

Remark 3.2 $s_{k_l}^{\text{new}}$ represents the computed minimum inter-event interval that guarantees convergence, and its derivation is based on system stability, which will be detailed in the subsequent proof.

We can compare the improved minimum inter-event interval expression (3.48) and the one presented in the previous section, i.e., expression (3.14). By comparing the left-hand sides of (3.14) and (3.48), it can be observed that the improved expression (3.48) no longer relies on the Lipschitz constant L_q, but instead applies a more refined formulation. Since the Lipschitz constant L_q is defined over the entire state space, it essentially represents the maximum possible value throughout the domain. As a result, the improved condition reduces conservatism and allows for a larger inter-event interval under the same conditions.

From the comparison of the right-hand sides of (3.14) and (3.48), it is evident that the improved expression replaces the single state and control term in (3.14) with multiple variable state and control. This modification effectively enlarges the feasible region of the inequality, thereby enabling a longer inter-event interval under equivalent conditions.

Although the improved expression (3.48) introduces a slight increase in computational complexity compared to (3.14), the additional computational burden is negligible in practice.

To ensure that the minimum inter-event interval satisfies $s_{k_l}^{\text{new}} \geq 1$, by substituting $z = 1$ into (3.48) and applying appropriate inequality relaxation, a sufficient condition guaranteeing the requirement can be derived as follows:

$$
\begin{aligned}
&\eta^2\bar{\lambda}(Q)\frac{(1+L_h)^{2(N_{k_0}-1)} - 1}{(1+L_h)^2 - 1} + 2\psi\eta\sqrt{\bar{\lambda}(Q)}\frac{(1+L_h)^{N_{k_0}-1} - 1}{L_h} \\
&+ \eta\sqrt{\bar{\lambda}(P)}(1+L_h)^{N_{k_0}-1}(r+\varepsilon) \leq (1-\sigma)\frac{\underline{\lambda}(Q)}{\bar{\lambda}(P)}r^2,
\end{aligned}
$$

where $\psi = \sup_{x\in\mathcal{X}} \|x\|_Q$.

After computing the minimum inter-event interval $s_{k_l}^{\text{new}}$, it is necessary to design the triggering mechanism. The design procedure follows a similar approach as previously described, where aperiodic sampling and detection are employed. At each sampling instant, the current state information is used in a feedback manner to compute the next sampling interval via an inverse-mapping method. For detailed procedures, refer to Algorithms 2 and 3.

The key difference from the previous algorithm lies in the fact that the prediction horizon N_{k_l} is no longer fixed but varies over time. Therefore, when calculating the triggering interval τ_{k_l}, the impact of the varying prediction horizon must be taken into account. Based on the above analysis, the triggering interval under the improved algorithm can be formulated as follows:

$$\tau_{k_l} = \begin{cases} \sum_{i=0}^{n} \gamma_l^i, & if\ \gamma_l^n = 0, \sum_{i=0}^{n} \gamma_l^i < N_{k_l}, \\ N_{k_l}, & if\ \sum_{i=0}^{n} \gamma_l^i \geq N_{k_l}, \end{cases} \tag{3.49}$$

where $\gamma_l^i = s_{k_l}^{\text{new}} - \left[\phi^{-1}(\left\|x(k_l^i) - \hat{x}^*(k_l^i\ |k_l)\right\|_P)\right]$ is the sampling interval. The sampling instant satisfies $k_l^{i+1} = k_l^i + \gamma_l^i$ and $k_l^0 = k_l$.

Using the improved triggering mechanism proposed in this section, the deviation between the predicted state and the actual state at the triggering instant will not exceed the desired triggering threshold $\phi(s_{k_l}^{\text{new}})$, which can be described by the following lemma.

Lemma 3.4 *Suppose that k_l is the previous triggering instant. If the improved aperiodic sampling triggering mechanism, including* (3.47) *and* (3.49)*, is applied, then at the next triggering instant, the following condition holds:*

$$\|x(k_{l+1}) - \hat{x}^*(k_{l+1}\ |k_l)\|_P \leq \phi(s_{k_l}^{\text{new}}), \tag{3.50}$$

where $k_{l+1} = k_l + \tau_{k_l}$ is the next triggering instant, and $\phi\left(s_{k_l}^{\text{new}}\right)$ is the desired triggering threshold.

The proof of Lemma 3.4 is basically similar to that of Lemma 3.3, so the detailed proof is omitted here.

3.3.2 Feasibility and Stability Concerning the Improved Aperiodic Triggering Mechanism

This section presents an analysis of the feasibility and stability of the improved triggering mechanism. Due to the relaxation of restrictive conditions in the improved triggering mechanism (specifically, the removal of the Lipschitz continuity assumption in Assumption (3.2) and the adoption of a predictive step-size updating strategy), an investigation into its feasibility and stability is necessary. Furthermore, in the stability proof provided herein, by imposing explicit bounds on the disturbances, it is shown that the system state enters a terminal invariant set. Compared with the result in Theorem 3.2, this section yields a stronger conclusion on the system stability.

First, the feasibility of the optimization problem is analyzed. Since the improved triggering mechanism incorporates a predictive step-size updating strategy, a slight modification to the original Assumption 3.4 is required. Accordingly, Assumption 3.4 is replaced with the newly introduced Assumption 3.5.

Assumption 3.5 The optimal control problem (3.41)–(3.46) is assumed to be feasible at the initial time instant k_0, and it holds that $\Omega_r \subseteq \mathbb{X} \ominus X_{N_{k_0}}$.

The feasibility theorem derived under the improved triggering mechanism is given below.

Theorem 3.3 Consider the actual system (3.1), and suppose that Assumptions 3.1, 3.3, and 3.5 hold. If the disturbance bounds and terminal constraints satisfy the following conditions:

$$\eta \leq \frac{(r-\varepsilon)L_h}{\sqrt{\bar{\lambda}(P)}(1+L_h)^{N_{k_0}}}, \quad 1-\frac{\underline{\lambda}(Q^*)}{\bar{\lambda}(P)} \leq \frac{\varepsilon^2}{r^2}, \tag{3.51}$$

then, under the improved aperiodic sampling-triggered mechanism, the optimization problem (3.41)–(3.46) is recursively feasible.

Proof: The feasibility is proved by using mathematical induction. According to Assumption 3.5, the optimal control problem is solvable at the initial time k_0.

Suppose the problem remains feasible at time k_l. To demonstrate the recursive feasibility of the optimization problem, it is necessary to ensure that the problem remains feasible at the subsequent update time k_{l+1}. In other words, we must show that, at time k_{l+1}, there exists at least one feasible control sequence for the optimization problem (3.41)–(3.46) such that the terminal constraint, state constraints, and control constraints prescribed therein are all satisfied.

Consider the case where the triggering interval satisfies $\tau_{k_l} < N_{k_l}$. Based on the optimal control sequence and the state feedback control law obtained at time k_l, a feasible control sequence at time k_{l+1} can be constructed by:

$$\tilde{u}(k_{l+1}+i\,|k_{l+1}) = \begin{cases} \hat{u}^*(k_{l+1}+i\,|k_l), & i \in [0, N_{k_l}-\tau_{k_l}-1], \\ K\tilde{x}(k_{l+1}+i\,|k_{l+1}), & i \in [N_{k_l}-\tau_{k_l}, N_{k_{l+1}}), \end{cases} \tag{3.52}$$

and the corresponding state sequence is $\tilde{x}(k_{l+1}+i\,|k_{l+1})$.

Consider the interval $[k_{l+1}, k_l + N_{k_l}]$. For each $i \in \mathbb{N}_{[0,N_{k_l}-\tau_{k_l}]}$, the deviation between the state corresponding to the feasible control sequence $\tilde{u}(k_{l+1}+i \mid k_{l+1})$ and the state corresponding to the optimal control sequence $\hat{\mathbf{u}}^*(k_l)$ satisfies:

$$\begin{aligned} \|\tilde{x}(k_{l+1}+i\,|k_{l+1}) - \hat{x}^*(k_{l+1}+i\,|k_l)\|_P &\leq \|\tilde{x}(k_{l+1}\,|k_{l+1}) - \hat{x}^*(k_{l+1}\,|k_l)\|_P \\ &\quad + \sum_{s=0}^{i-1} L_h \|\tilde{x}(k_{l+1}+s\,|k_{l+1}) - \hat{x}^*(k_{l+1}+s\,|k_l)\|_P. \end{aligned} \tag{3.53}$$

Since $\tilde{x}(k_{l+1} \mid k_{l+1}) = x(k_{l+1})$, and by invoking Lemma 3.4, it holds that:

$$\|\tilde{x}(k_{l+1} \mid k_{l+1}) - \hat{x}^*(k_{l+1} \mid k_l)\|_P \leq \phi(s_{k_l}^{\text{new}}).$$

Substituting the above inequality into (3.53) yields:

$$\|\tilde{x}(k_{l+1} + i \mid k_{l+1}) - \hat{x}^*(k_{l+1} + i \mid k_l)\|_P \leq \eta\sqrt{\overline{\lambda}(P)}\frac{(1+L_h)^{s_{k_l}^{\text{new}}} - 1}{L_h}(1+L_h)^i. \tag{3.54}$$

Substituting $i = N_{k_l} - \tau_{k_l}$ into (3.54) yields:

$$\|\tilde{x}(k_l + N_{k_l} \mid k_{l+1})\|_P \leq \|\hat{x}^*(k_l + N_{k_l} \mid k_l)\|_P + \eta\sqrt{\overline{\lambda}(P)}\frac{(1+L_h)^{s_{k_l}^{\text{new}}} - 1}{L_h}(1+L_h)^{N_{k_l}-\tau_{k_l}}. \tag{3.55}$$

According to the improved triggering mechanism, it holds that $N_{k_0} \geq N_{k_l} > \tau_{k_l} \geq s_{k_l}^{\text{new}}$. Moreover, from (3.46), it follows that $\|\hat{x}^*(k_l + N_{k_l} \mid k_l)\|_P \leq \varepsilon$. Together with the disturbance satisfying condition (3.51), substituting into (3.55) yields:

$$\|\tilde{x}(k_l + N_{k_l} \mid k_{l+1})\|_P \leq \varepsilon + r - \varepsilon \leq r, \tag{3.56}$$

indicating that $\tilde{x}(k_l + N_{k_l} \mid k_{l+1}) \in \Omega_r$.

Consider the interval $[k_l + N_{k_l}, k_{l+1} + N_{k_{l+1}}]$. It follows from Assumption 3.3 that:

$$\begin{aligned}
&V_f(\tilde{x}(k_l + N_{k_l} + i + 1 \mid k_{l+1})) - V_f(\tilde{x}(k_l + N_{k_l} + i \mid k_{l+1})) \\
&\leq -\|\tilde{x}(k_l + N_{k_l} + i \mid k_{l+1})\|_{Q^*}^2 \\
&\leq -\frac{\underline{\lambda}(Q^*)}{\overline{\lambda}(P)} V_f(\tilde{x}(k_l + N_{k_l} + i \mid k_{l+1})).
\end{aligned} \tag{3.57}$$

Substituting $i = k_{l+1} + N_{k_{l+1}} - k_l - N_{k_l} - 1$ into (3.57) and applying the recursive relation yields:

$$\begin{aligned}
V_f(\tilde{x}(k_{l+1} + N_{k_{l+1}} \mid k_{l+1})) &\leq \left(1 - \frac{\underline{\lambda}(Q^*)}{\overline{\lambda}(P)}\right) V_f(\tilde{x}(k_{l+1} + N_{k_{l+1}} - 1 \mid k_{l+1})) \\
&\leq \left(1 - \frac{\underline{\lambda}(Q^*)}{\overline{\lambda}(P)}\right)^2 V_f(\tilde{x}(k_{l+1} + N_{k_{l+1}} - 2 \mid k_{l+1})) \\
&\cdots\cdots \\
&\leq \left(1 - \frac{\underline{\lambda}(Q^*)}{\overline{\lambda}(P)}\right)^{\tau_{k_l} - \lfloor \alpha\tau_{k_l} \rfloor + 1} V_f(\tilde{x}(k_l + N_{k_l} \mid k_{l+1})).
\end{aligned} \tag{3.58}$$

Substituting (3.51) and (3.56) into (3.58), and utilizing the inequality $\tau_{k_l} - \lfloor \alpha \tau_{k_l} \rfloor \geq 0$ yields:

$$V_f(\tilde{x}(k_{l+1} + N_{k_{l+1}} | k_{l+1})) \leq \left(1 - \frac{\underline{\lambda}(Q^*)}{\bar{\lambda}(P)}\right) V_f(\tilde{x}(k_l + N_{k_l} | k_{l+1})) \leq \varepsilon^2, \quad (3.59)$$

implying that $\tilde{x}(k_{l+1} + N_{k_{l+1}} | k_{l+1}) \in \Omega_\varepsilon$, i.e., the terminal constraint is satisfied.

Consider the state constraints. For $i \in \mathbb{N}_{[0, N_{k_l} - \tau_{k_l}]}$, it follows from (3.54) that:

$$\|\tilde{x}(k_{l+1} + i | k_{l+1})\|_P \leq \|\hat{x}^*(k_{l+1} + i | k_l)\|_P + \eta\sqrt{\bar{\lambda}(P)}\frac{(1 + L_h)^{s_{k_l}^{\text{new}}} - 1}{L_h}(1 + L_h)^i. \quad (3.60)$$

With $\hat{x}^*(k_{l+1} + i | k_l) \in \mathbb{X} \ominus X_{i+\tau_{k_l}}$, it follows from (3.60) that:

$$\tilde{x}(k_{l+1} + i | k_{l+1}) \in \mathbb{X} \ominus X_{i+\tau_{k_l}} \oplus B\left(\eta\sqrt{\bar{\lambda}(P)}\frac{(1 + L_h)^{s_{k_l}^{\text{new}}} - 1}{L_h}(1 + L_h)^i\right), \quad (3.61)$$

where $B(s) \triangleq \{x : \|x\|_P \leq s\}$.

It then follows from the triggering mechanism that $\tau_{k_l} \geq s_{k_l}^{\text{new}}$, leading to:

$$X_i \subseteq X_{i+\tau_{k_l}} \ominus B\left(\eta\sqrt{\bar{\lambda}(P)}\frac{(1 + L_h)^{s_{k_l}^{\text{new}}} - 1}{L_h}(1 + L_h)^i\right). \quad (3.62)$$

Substituting (3.62) into (3.61) yields:

$$\tilde{x}(k_{l+1} + i | k_{l+1}) \in \mathbb{X} \ominus X_i, \quad (3.63)$$

implying that, for $i \in \mathbb{N}_{[0, N_{k_l} - \tau_{k_l}]}$, state constraints are all satisfied.

For each $i \in \mathbb{N}_{[N_{k_l} - \tau_{k_l} + 1, N_{k_{l+1}}]}$, it follows from (3.56) that:

$$\tilde{x}(k_l + N_{k_l} | k_{l+1}) \in \Omega_r.$$

According to Assumption 3.3, under state feedback control law:

$$\tilde{u}(k_{l+1} + i | k_{l+1}) = K\tilde{x}(k_{l+1} + i | k_{l+1}),$$

the predicted state satisfies $\tilde{x}(k_{l+1} + i | k_{l+1}) \in \Omega_r$. Moreover, by Assumption 3.5, we have $\Omega_r \subseteq \mathbb{X} \ominus X_{N_{k_0}}$. Therefore, for all $i \in \mathbb{N}_{[N_{k_l} - \tau_{k_l} + 1, N_{k_{l+1}}]}$, the predicted state satisfies the constraint:

$$\tilde{x}(k_{l+1} + i | k_{l+1}) \in \mathbb{X} \ominus X_{N_{k_0}} \subseteq \mathbb{X} \ominus X_i. \quad (3.64)$$

Combining (3.63) and (3.64), it follows that for all $i \in \mathbb{N}_{[0,N_{k_{l+1}}]}$, the state constraint $\tilde{x}(k_{l+1} + i \mid k_{l+1}) \in \mathbb{X} \ominus X_i$ is satisfied.

Based on the construction procedure of the feasible control sequence (3.52) and Assumption 3.3, it follows that the control constraints are satisfied at all sampling times.

If the triggering interval satisfies $\tau_{k_l} = N_{k_l}$, then a feasible control sequence can be constructed by $\tilde{u}(k_{l+1} + i \mid k_{l+1}) = K\tilde{x}(k_{l+1} + i \mid k_{l+1})$, for $i \in [0, N_{k_{l+1}})$. As a special case of (3.52), it can be shown, by using the same reasoning, that all constraints are satisfied.

In summary, at time k_{l+1}, the constructed feasible control sequence ensures satisfaction of the terminal constraint, state constraints, and control constraints. Therefore, the optimization problem is recursively feasible.

Remark 3.3 Since $\tau_{k_l} - \lfloor \alpha\tau_{k_l} \rfloor \geq 0$ holds for all $\alpha \in (0, 1]$, it follows from (3.58) and (3.59) that the terminal constraint is satisfied. This indicates that α does not affects the feasibility of the optimization problem. However, an analysis of (3.47), in conjunction with (3.48), reveals that different values of α influence the prediction horizon N_{k_l}, thereby affecting the minimum triggering interval $s_{k_l}^{\text{new}}$.

The following analysis investigates the stability of the closed-loop system under the improved triggering mechanism.

Theorem 3.4 If Assumptions 3.1, 3.3, and 3.5 hold, and the conditions stated in Theorem 3.3 are satisfied, then under the improved aperiodic triggering mechanism, the system state converges to the set Ω_r within a finite time. Furthermore, if the disturbance satisfies the following condition:

$$2\eta\delta r + \xi\eta^2 + \frac{2\,\|P\|\,\eta}{\sqrt{\underline{\lambda}(P)}}\sqrt{2\eta\delta r + \eta^2\xi + \eta^2\bar{\lambda}(P)} \leq r^2, \tag{3.65}$$

where

$$\delta \triangleq \frac{\bar{\lambda}(P)\,\|P\|}{\underline{\lambda}(Q^*)\sqrt{\underline{\lambda}(P)}}, \quad \xi \triangleq \frac{\bar{\lambda}^2(P)}{\underline{\lambda}(Q^*)},$$

then Ω_r is a robust invariant set, and the system state remains within this set thereafter.

Proof: Suppose that at the initial time k_0, the initial state lies outside the set Ω_r.

Define the cost function associated with the optimal control sequence $\hat{\mathbf{u}}^*(k_l)$ at time $x(k_l)$ as a Lyapunov function candidate, denoted by:

$$V(x(k_l)) = J(x(k_l), \hat{\mathbf{u}}^*(k_l)).$$

Correspondingly, let the cost function evaluated at time $x(k_{l+1})$ under a feasible control sequence be represented by:

$$\tilde{V}(x(k_{l+1})) = J(x(k_{l+1}), \tilde{u}(k_{l+1})).$$

By the optimality of the cost function, we have $V(x(k_{l+1})) \leq \tilde{V}(x(k_{l+1}))$. Consequently, the difference between $V(x(k_{l+1}))$ and $V(x(k_l))$ satisfies:

$$\begin{aligned} V(x(k_{l+1})) - V(x(k_l)) &\leq J(x(k_{l+1}), \tilde{u}(k_{l+1})) - J(x(k_l), \hat{\mathbf{u}}^*(k_l)) \\ &\leq \Delta V_1 + \Delta V_2 + \Delta V_3, \end{aligned} \tag{3.66}$$

where

$$\begin{aligned} \Delta V_1 &= -\sum_{i=0}^{\tau_{k_l}-1} (\|\hat{x}^*(k_l+i\,|k_l)\|_Q^2 + \|\hat{u}^*(k_l+i\,|k_l)\|_R^2), \\ \Delta V_2 &= \sum_{i=\tau_{k_l}}^{N_{k_l}-1} (\|\tilde{x}(k_l+i\,|k_{l+1})\|_Q^2 - \|\hat{x}^*(k_l+i\,|k_l)\|_Q^2), \\ \Delta V_3 &= \sum_{i=N_{k_l}}^{N_{k_{l+1}}+\tau_{k_l}-1} (\|\tilde{x}(k_l+i\,|k_{l+1})\|_Q^2 + \|\tilde{u}(k_l+i\,|k_{l+1})\|_R^2) \\ &\quad + \left\|\tilde{x}(k_{l+1}+N_{k_{l+1}}\,|k_{l+1})\right\|_P^2 - \left\|\hat{x}^*(k_l+N_{k_l}\,|k_l)\right\|_P^2. \end{aligned}$$

For ΔV_2, it can be calculated by:

$$\begin{aligned} \Delta V_2 &\leq \sum_{i=\tau_{k_l}}^{N_{k_l}-1} (\|\tilde{x}(k_l+i\,|k_{l+1}) - \hat{x}^*(k_l+i\,|k_l)\|_Q)(\|\tilde{x}(k_l+i\,|k_{l+1})\|_Q \\ &\quad + \|\hat{x}^*(k_l+i\,|k_l)\|_Q) \\ &\leq \sum_{i=\tau_{k_l}}^{N_{k_l}-1} \eta\sqrt{\bar{\lambda}(Q)}\frac{(1+L_h)^{s_{k_l}^{\text{new}}}-1}{L_h(1+L_h)^{\tau_{k_l}}}(1+L_h)^i \\ &\quad \left(\eta\sqrt{\bar{\lambda}(Q)}\frac{(1+L_h)^{s_{k_l}^{\text{new}}}-1}{L_h(1+L_h)^{\tau_{k_l}}}(1+L_h)^i + 2\,\|\hat{x}^*(k_l+i\,|k_l)\|_Q\right) \\ &\leq \sum_{i=s_{k_l}^{\text{new}}}^{N_{k_l}-1} \eta\sqrt{\bar{\lambda}(Q)}\frac{(1+L_h)^{s_{k_l}^{\text{new}}}-1}{L_h(1+L_h)^{s_{k_l}^{\text{new}}}}(1+L_h)^i \\ &\quad \left(\eta\sqrt{\bar{\lambda}(Q)}\frac{(1+L_h)^{s_{k_l}^{\text{new}}}-1}{L_h(1+L_h)^{s_{k_l}^{\text{new}}}}(1+L_h)^i + 2\,\|\hat{x}^*(k_l+i\,|k_l)\|_Q\right) \\ &\leq \sum_{i=s_{k_l}^{\text{new}}}^{N_{k_l}-1} (\beta^2(i) + 2\beta(i)\,\|\hat{x}^*(k_l+i\,|k_l)\|_Q), \end{aligned} \tag{3.67}$$

where

$$\beta(i) = \eta\sqrt{\bar{\lambda}(Q)}\frac{(1+L_h)^{s_{k_l}^{\text{new}}} - 1}{L_h(1+L_h)^{s_{k_l}^{\text{new}}}}(1+L_h)^i.$$

It then follows from Assumption 3.3 that:

$$\begin{aligned} &\left\|\tilde{x}(k_l + N_{k_l} + i + 1 \,|k_{l+1})\right\|_P^2 - \left\|\tilde{x}(k_l + N_{k_l} + i \,|k_{l+1})\right\|_P^2 \\ &\leq -\left\|\tilde{x}(k_l + N_{k_l} + i \,|k_{l+1})\right\|_Q^2 - \left\|\tilde{u}(k_l + N_{k_l} + i \,|k_{l+1})\right\|_R^2. \end{aligned} \tag{3.68}$$

By summing (3.68) from $i = 0$ to $i = \tau_{k_l} + N_{k_{l+1}} - N_{k_l} - 1$, we obtain:

$$\begin{aligned} &\left\|\tilde{x}(k_{l+1} + N_{k_{l+1}} \,|k_{l+1})\right\|_P^2 - \left\|\tilde{x}(k_l + N_{k_l} \,|k_{l+1})\right\|_P^2 \\ &\leq -\sum_{i=N_{k_l}}^{N_{k_{l+1}}+\tau_{k_l}-1} (\|\tilde{x}(k_l + i \,|k_{l+1})\|_Q^2 + \|\tilde{u}(k_l + i \,|k_{l+1})\|_R^2). \end{aligned} \tag{3.69}$$

Substituting (3.69) into ΔV_3 yields:

$$\begin{aligned} \Delta V_3 &\leq \left\|\tilde{x}(k_l + N_{k_l} \,|k_{l+1})\right\|_P^2 - \left\|\hat{x}^*(k_l + N_{k_l} \,|k_l)\right\|_P^2 \\ &\leq \left(\left\|\tilde{x}(k_l + N_{k_l} \,|k_{l+1})\right\|_P - \left\|\hat{x}^*(k_l + N_{k_l} \,|k_l)\right\|_P\right) \\ &\times \left(\left\|\tilde{x}(k_l + N_{k_l} \,|k_{l+1})\right\|_P + \left\|\hat{x}^*(k_l + N_{k_l} \,|k_l)\right\|_P\right). \end{aligned} \tag{3.70}$$

Substituting (3.55), $\left\|\tilde{x}(k_l + N_{k_l} \,|k_{l+1})\right\|_P \leq r$, and $\left\|\hat{x}^*(k_l + N_{k_l} \,|k_l)\right\|_P \leq \varepsilon$ into (3.70) yields:

$$\Delta V_3 \leq \eta\sqrt{\bar{\lambda}(P)}\frac{(1+L_h)^{N_{k_l}} - (1+L_h)^{N_{k_l}-s_{k_l}^{\text{new}}}}{L_h}(r+\varepsilon). \tag{3.71}$$

Substituting (3.67) and (3.71) into (3.66), and incorporating the triggering interval (3.48) yields:

$$\begin{aligned} V(x(k_{l+1})) - V(x(k_l)) &\leq -\sigma\sum_{i=0}^{\tau_{k_l}-1}\left(\|\hat{x}^*(k_l + i \,|k_l)\|_Q^2 + \|\hat{u}^*(k_l + i \,|k_l)\|_R^2\right) \\ &\leq -\sigma\frac{\underline{\lambda}(Q)}{\bar{\lambda}(P)}\varepsilon^2. \end{aligned} \tag{3.72}$$

Applying summation to (3.72) yields:

$$V(x(k_{l+1})) \leq V(x(k_0)) - (l+1)\sigma\frac{\underline{\lambda}(Q)}{\bar{\lambda}(P)}\varepsilon^2. \tag{3.73}$$

Suppose that $l \to \infty$. Then the right-hand side of (3.73) tends to $-\infty$. However, by definition, the left-hand side of (3.73) is nonnegative. Therefore, by contradiction, l must be bounded. As a result, the system state will enter the set Ω_r within a finite time.

Once the system state enters the terminal set Ω_r, the actual system begins to operate under the state feedback control law $u(k) = Kx(k)$. Taking $V_f(x(k)) = \|x(k)\|_P^2$ as a Lyapunov function candidate and applying Assumption 3.3, it follows that:

$$
\begin{aligned}
V_f(x(k+1)) - V_f(x(k)) &= \|f(x(k), Kx(k)) + w(k)\|_P^2 - \|x(k)\|_P^2 \\
&\leq -\|x(k)\|_{Q^*}^2 + 2w^T(k)Pf(x(k), Kx(k)) + \|w(k)\|_P^2 \\
&\leq -\frac{\underline{\lambda}(Q^*)}{\bar{\lambda}(P)}\|x(k)\|_P^2 + \frac{2\eta r\ \|P\|}{\sqrt{\underline{\lambda}(P)}} + \bar{\lambda}(P)\eta^2 \\
&\leq -\frac{\underline{\lambda}(Q^*)}{\bar{\lambda}(P)}(\|x(k)\|_P^2 - 2\delta\eta r - \xi\eta^2),
\end{aligned}
\tag{3.74}
$$

where

$$
\delta = \frac{\|P\|\ \bar{\lambda}(P)}{\sqrt{\underline{\lambda}(P)}\underline{\lambda}(Q^*)}, \quad \xi = \frac{\bar{\lambda}^2(P)}{\underline{\lambda}(Q^*)}.
$$

From (3.74), it follows that the system state converges to the set $\Omega_\eta = \{x \mid \|x\|_P^2 \leq 2\delta\eta r + \xi\eta^2\}$.

If the system state enters the set Ω_η at time k_p, then at the next sampling instant, the system state satisfies:

$$
\begin{aligned}
\|x(k_p+1)\|_P^2 &= \|f(x(k_p), Kx(k_p)) + w(k_p)\|_P^2 \\
&\leq 2\delta\eta r + \xi\eta^2 + \frac{2\eta\ \|P\|}{\sqrt{\underline{\lambda}(P)}}\sqrt{2\delta\eta r + \xi\eta^2} + \bar{\lambda}(P)\eta^2.
\end{aligned}
\tag{3.75}
$$

If the disturbance satisfies condition (3.65), substituting it into (3.75) yields:

$$
\left\|x(k_p+1)\right\|_P^2 \leq r^2,
$$

indicating that, at time $k_p + 1$, the system state lies within the set Ω_r.

At time $k_p + 1$, two cases need to be considered. In the first case, the state lies within the set Ω_η. In this situation, using an approach similar to that in (3.75), it can be shown that, at time $k_p + 2$, the system state lies within the set Ω_r.

In the second case, the state at time $k_p + 1$ lies in the set $\Omega_r \backslash \Omega_\eta$, i.e., within Ω_r but outside of Ω_η. In this case, according to (3.74), the Lyapunov function candidate is decreasing, and thus the system state remains within the set Ω_r at time $k_p + 2$.

Since these two cases repeat indefinitely in subsequent time steps, the recursive argument shows that the state never leaves the set Ω_r.

If the disturbance satisfies (3.65), then the set Ω_η is a subset of Ω_r, and the set Ω_r can be partitioned into two regions: $\Omega_r \backslash \Omega_\eta$ and Ω_η. If the system state lies within $\Omega_r \backslash \Omega_\eta$, then according to (3.74), the state converges to the set Ω_η.

Due to the influence of disturbances, the state trajectory within Ω_η may potentially leave Ω_η. Even in the worst-case scenario where the state exits Ω_η, condition (3.65) guarantees that the state remains within $\Omega_r \backslash \Omega_\eta$. Hence, the system state always stays within the set Ω_r, and therefore, Ω_r is a robust invariant set. This completes the proof. □

Remark 3.4 According to (3.72), different values of σ influence the decay rate of the Lyapunov function $V(x(k_l))$. Moreover, (3.48) indicates that the choice of σ also affects the minimum triggering interval $s_{k_l}^{\text{new}}$.

3.4 A Simulation Example

Consider a simplified mass–spring–damping structure in the car suspension system, as illustrated by Figure 3.3. It can be described by the following model:

$$\begin{cases} x_1(k+1) = x_1(k) + \mu x_2(k) \\ x_2(k+1) = x_2(k) + \mu(-\dfrac{\xi}{M_m} e^{-x_1(k)} x_1(k) - \dfrac{h_d}{M_m} x_2(k) + \dfrac{u(k)}{M_m} + w_m(k)), \end{cases} \tag{3.76}$$

where $x_1(k)$ and $x_2(k)$ denote the position and velocity of the mass, respectively. The discretization interval is set to $\mu = 0.1$; $w_m(k) \leq 0.0013$ denotes the bounded external disturbances, and it is generated using random values during simulation.

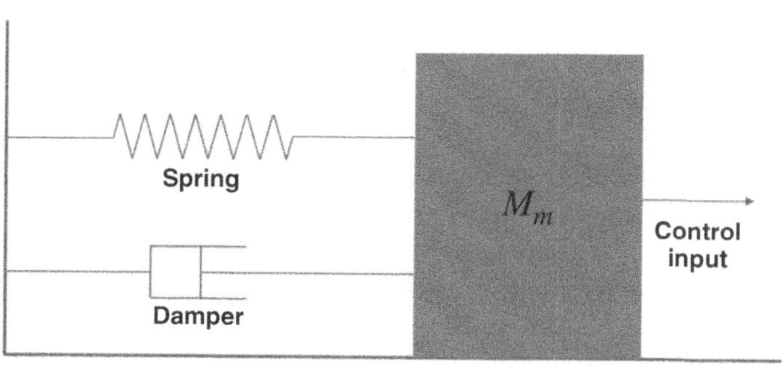

Figure 3.3 Simplified damping spring structure in a car suspension system.

The mass M_m is supposed to be 1.25 kg, and the spring stiffness is $\xi = 0.9$ N/m. The damping coefficient of the system is $h_d = 0.42$ N · s/m. The control input satisfies its constraints $-1\ \text{N} \leq u(k) \leq 1\ \text{N}$, and the state constraints are supposed to be $\mathbb{X} = \{x \mid \|x_1\| \leq 0.5\ \text{m},\ \ \|x_2\| \leq 1.5\ \text{m/s}\}$. According to [73], the Lipschitz constant can be computed by $L_h = 1.145$.

The weighting matrices are set to $Q = \text{diag}\{0.06, 0.06\}$ and $R = 0.001$. According to [74], the terminal weighting matrix can be calculated to be

$$P = \begin{bmatrix} 0.795 & 0.134 \\ 0.134 & 0.154 \end{bmatrix},$$

and the corresponding state feedback gain is $K = [-4.66, -6.45]$. Furthermore, the terminal set parameters are chosen by $r = 0.086$ and $\varepsilon = 0.082$ to ensure feasibility and stability. The associated sets are computed to be $\Omega_r = \{x \mid \|x\|P^2 \leq 0.0074\}$ and $\Omega\varepsilon = \{x \mid \|x\|_P^2 \leq 0.00672\}$. The initial state is assumed to be $x(k_0) = [0.5,\ -1.2]^T$.

The simulation example was processed using MATLAB [79] on a system equipped with an Intel® Core™ i5-10400 CPU. The initial prediction horizon was set to $N_{k_0} = 28$ to ensure the feasibility of the initial optimization problem. The contraction factor was chosen by $\alpha = 0.2$, and the performance factor was set to $\sigma = 0.4$.

Using the improved triggering strategy, the corresponding state and control trajectories are plotted and compared with those obtained under the self-triggered mechanism, as shown in Figure 3.4. In the self-triggered MPC framework, the next triggering instant is estimated based on the measurement taken at the previous triggering moment, eliminating the need for continuous state sampling. Analysis of Figure 3.4 reveals that the closed-loop system states converge to the set Ω_r, and the control constraints are satisfied throughout the entire process. As illustrated in Figure 3.4, the trajectories under the improved triggering mechanism and the self-triggered mechanism are nearly identical, with only minor differences. To quantitatively evaluate the system performance under different strategies, a performance index is constructed by:

$$J_{\text{per}} = \sum_{k=0}^{60} \left(\|x(k)\|_Q^2 + \|u(k)\|_R^2\right),$$

and computed accordingly. For the triggering mechanism proposed in this chapter, the performance index yields a value of 0.4949, whereas the self-triggered mechanism results in a slightly lower value of 0.4947. This indicates that the system performance under the proposed mechanism is marginally inferior to that of the self-triggered approach.

To highlight the advantages of the method proposed in this chapter, Figure 3.5 presents the sampling and computation instants corresponding to the proposed

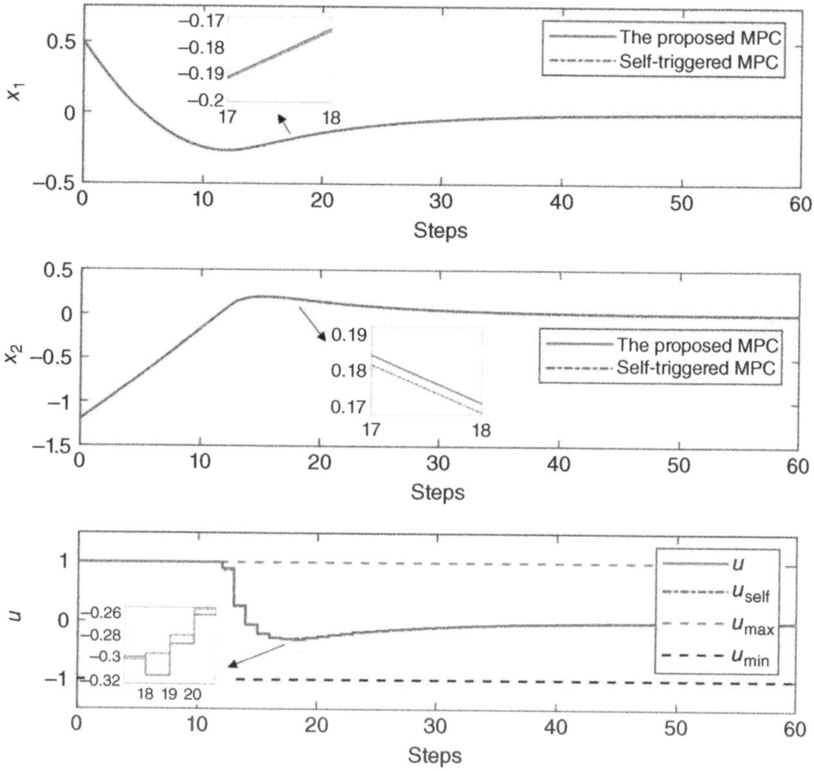

Figure 3.4 State and control trajectories under the aperiodic triggering mechanism and the self-triggered mechanism.

triggering mechanism and the self-triggered MPC approach described in [61]. The explanation of each type of instant is summarized in Table 3.1. In Figure 3.5, solid markers indicate computation instants: circular and diamond-shaped solid dots correspond to the instants when the optimization problem is solved, while asterisks denote local state feedback instants. Hollow dots represent sampling instants, at which the sensor merely acquires the system state without solving the optimization problem. During the initial phase, the sensor samples the system state at designated instants, and the nonlinear optimization problem is solved at the triggering instants. Once the system state enters the terminal set, a local state feedback control law is applied, and no further optimization is processed.

As shown in Figure 3.5, the number of optimization computations under the improved triggering mechanism is 4, whereas it is 17 under the self-triggered MPC strategy. This demonstrates that the proposed mechanism significantly reduces the frequency of solving optimization problems compared to the

Table 3.1 Description of sampling and computation instants in the discrete-time system.

Instant	Description
Sampling	Measure the system state, and check triggering condition.
Optimization	Measure the system state, and process optimization.
Feedback	Measure the system state, and calculate state feedback.

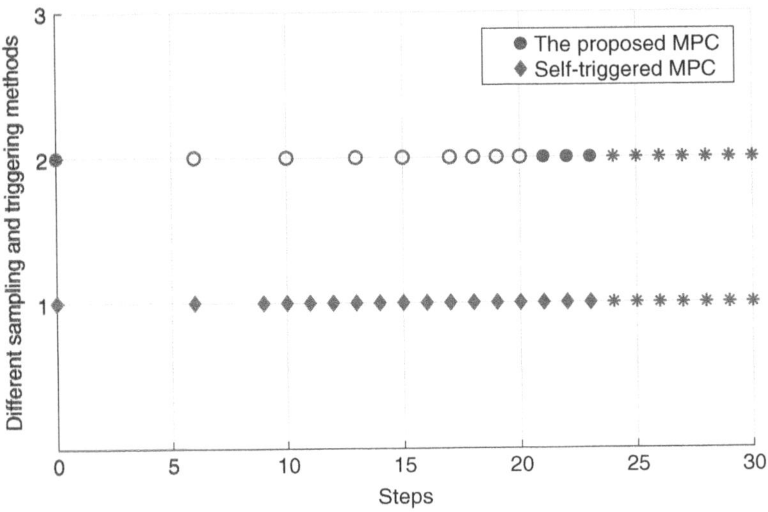

Figure 3.5 Sampling and computation instants corresponding to the aperiodic-triggered MPC and the self-triggered MPC.

self-triggered MPC. Furthermore, after repeating the simulation ten times and comparing the average simulation times, the proposed triggering mechanism yields an average computation time of 5.42 seconds, while the self-triggered MPC approach requires an average of 17.23 seconds. This indicates that the proposed strategy reduces computation time, thereby conserving computational resources and alleviating the processing burden on the CPU.

Selecting different design parameters affects the closed-loop performance. Figure 3.6 illustrates the system states and control trajectories under various initial prediction horizons. As shown in the figure, a shorter initial prediction horizon results in relatively faster state convergence but leads to more drastic variations in the control input. In contrast, a longer initial prediction horizon yields slower state convergence while producing smoother control input changes.

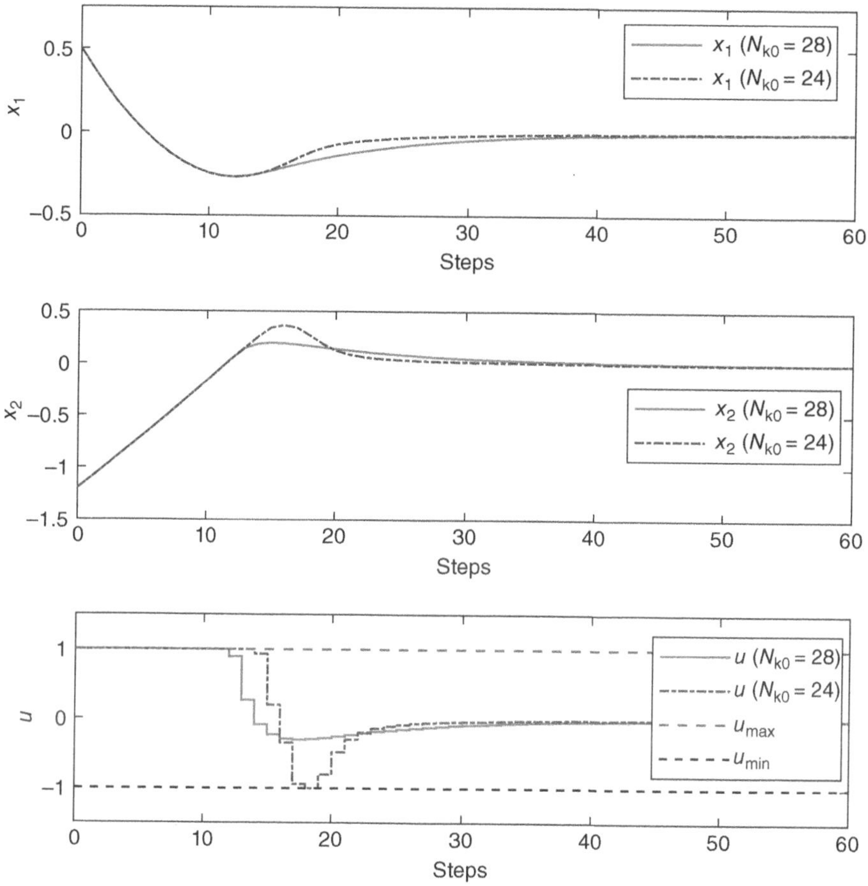

Figure 3.6 The system states and control trajectories corresponding to different initial prediction horizons.

Selection of the weight matrices depends on the required system performance. If the coefficients in the weight matrix Q are large, the system states converge with a relatively large rate. Conversely, if the coefficients in the weight matrix R are large, the control input converges with a relatively large rate. The design of the terminal weight matrix P depends on Q and R, and its design criteria should satisfy Assumption 3.3.

We compare the system performance under two different sets of weight matrix coefficients and plot them in Figure 3.7. The two sets of weight matrices are given

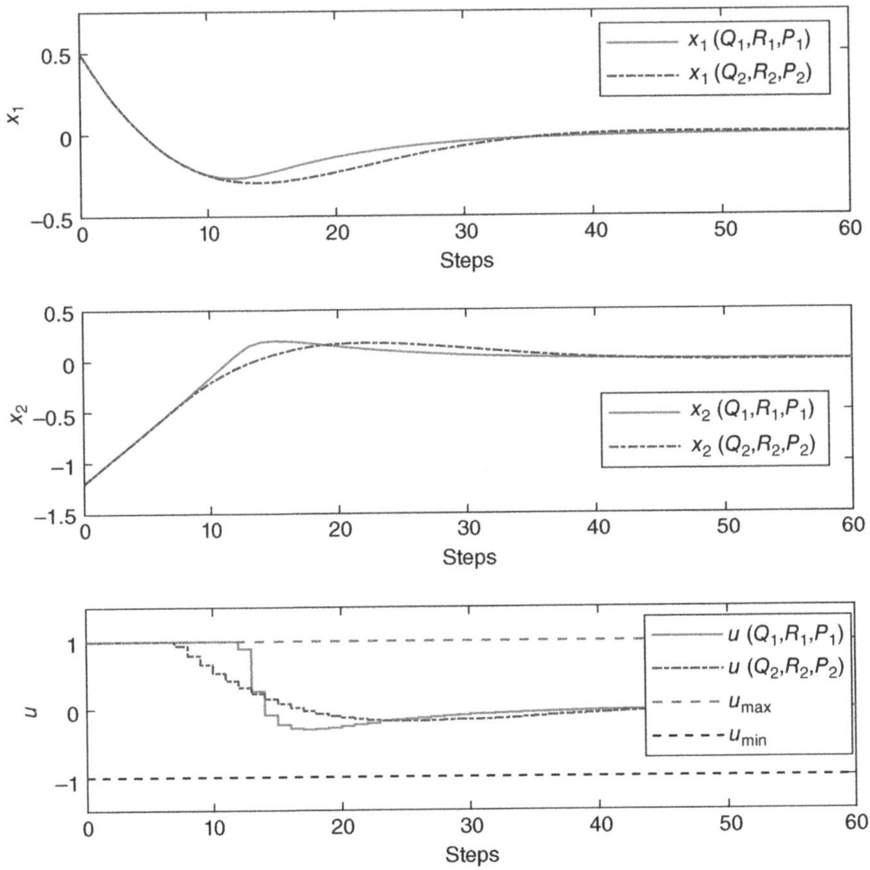

Figure 3.7 System states and control trajectories under different weighting coefficients.

by $Q_1 = \text{diag}\{0.06, 0.06\}$, $R_1 = 0.001$, and $Q_2 = \text{diag}\{0.02, 0.02\}$, $R_2 = 0.008$. The corresponding terminal weight matrices are calculated by:

$$P_1 = \begin{bmatrix} 0.795 & 0.134 \\ 0.134 & 0.154 \end{bmatrix}, \text{ and } P_2 = \begin{bmatrix} 0.814 & 0.126 \\ 0.126 & 0.392 \end{bmatrix}$$

From Figure 3.7, it can be seen that, if the state weight decreases, the convergence speed of the system states slows down. Conversely, when the control weight increases, the control input converges faster.

Due to the conservative nature of the scaling in the theoretical derivation, the actual system can maintain robust stability even under larger disturbances. Figure 3.8 illustrates the system state and control trajectories when the disturbance range satisfies $w_m(k) \leq 0.03$. From Figure 3.8, it can be witnessed that, even with

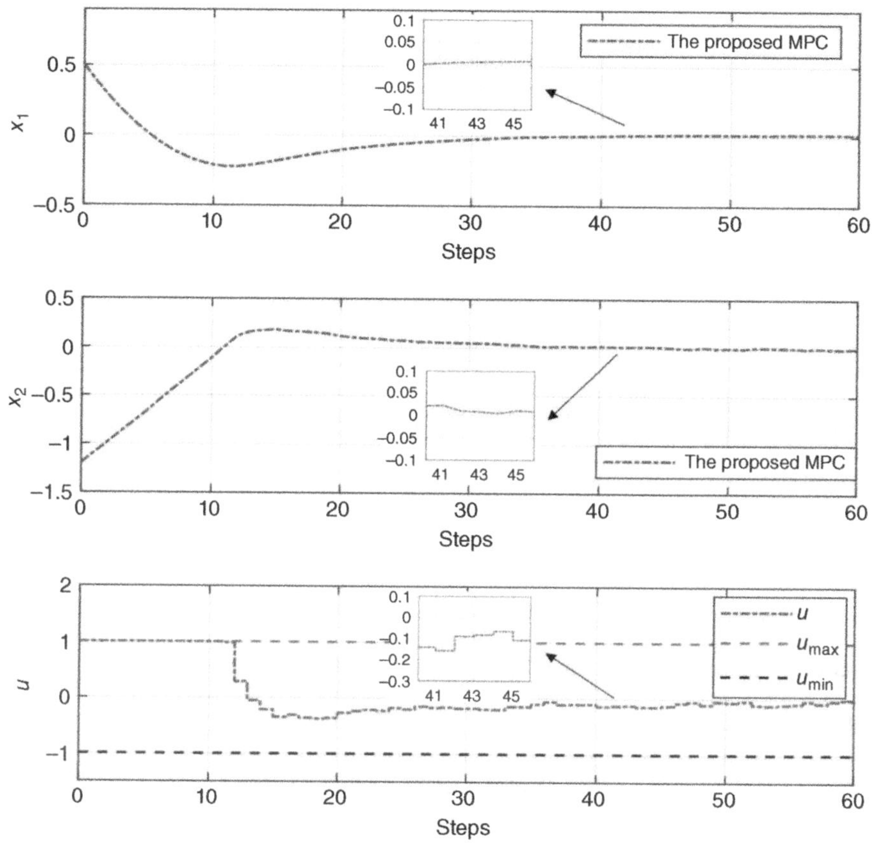

Figure 3.8 State and control signal trajectories under large disturbances.

significant disturbances, the system states still converge to the set Ω_r; and the control input converges to the small neighborhood of the origin, oscillating around the origin due to the disturbance influence.

The sampling and triggering instants under different values of the performance factor σ are compared, and the corresponding results are illustrated in Figure 3.9, where the performance factors are chosen by $\sigma = 0.3,\ 0.4,\ 0.5$, respectively. From Figure 3.9, it can be observed that as the value of the performance factor σ decreases, both the sampling interval and the triggering interval increase. Moreover, a smaller performance factor implies a slower convergence rate, potentially requiring more steps for the state to reach the terminal set. Conversely, as the performance factor σ increases, the sampling and triggering intervals decrease, and fewer steps are needed for the state to converge into the terminal set.

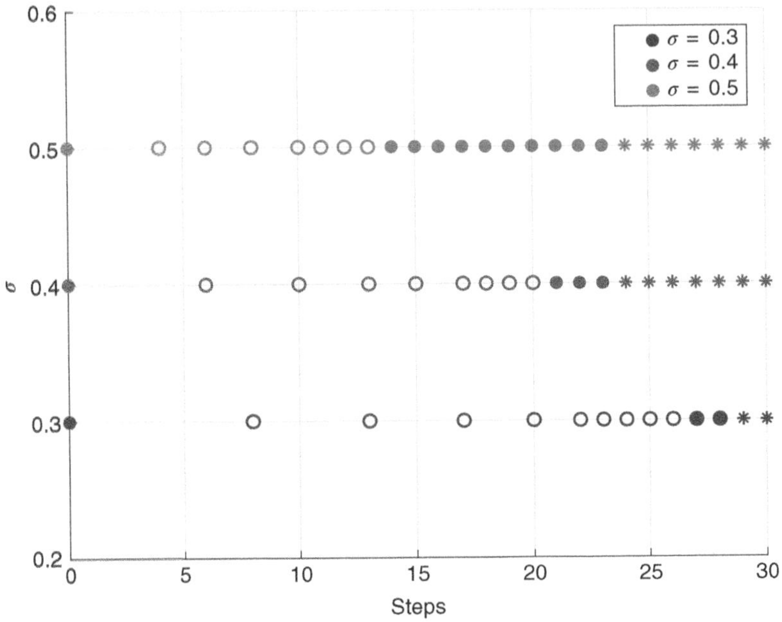

Figure 3.9 Sampling and computation times under different performance factors.

Table 3.2 Variation of prediction horizon under different contraction factors.

	Variation of the predictive horizon			
Values of contraction factor	**1st**	**2nd**	**3rd**	**4th**
$\alpha = 0.1$	28	27	27	27
$\alpha = 0.2$	28	26	25	25
$\alpha = 0.3$	28	25	24	23

To demonstrate the effectiveness of the prediction horizon updating strategy, Table 3.2 presents the variation in prediction horizon under different contraction factors α, with the initial prediction horizon set to 28 steps. From Table 3.2, it can be observed that updating the prediction horizon using the contraction factor α leads to a progressive reduction in the required prediction steps for solving the optimization problem, thereby alleviating the computational burden and conserving computational resources.

3.5 Conclusion

This chapter addresses the control problem for discrete systems subject to disturbances, state constraints, and control constraints. It proposes an event-based aperiodic sampling MPC method. The core of this strategy lies in designing a triggering threshold based on Lyapunov convergence properties. The method calculates the time interval by considering the worst-case impact of potential disturbances, using this time as the sampling instant. Based on the deviation between the actual and predicted states at this new sampling instant, a new sampling interval required to reach the triggering threshold is recalculated. This process is repeated until the calculated sampling interval becomes zero, and this final sampling instant is then set as the new triggering time. Through state feedback information at each sampling, this approach achieves event-based aperiodic sampling with a finite number of state samplings and detections.

Furthermore, based on the predictive nature of MPC, this chapter introduces a predictive horizon update strategy. This strategy reduces the number of prediction steps as the state approaches the target, thereby decreasing the computational burden of the optimization problem. The recursive feasibility of the optimization problem and the stability of the closed-loop system, when using this aperiodic sampling strategy, are proven in this chapter. Finally, simulation experiments demonstrate that the proposed strategy reduces the conservatism of the self-triggered estimation process, lowers the frequency of solving the optimization problem compared to traditional methods, saves computational resources, and improves the practical applicability of the algorithm.

4

Composite Event-triggered MPC based on Disturbance Compensation

Model predictive control (MPC) relies on a system model to predict future states and optimize a cost function. This inherently demands high accuracy from the system model itself. However, in real-world applications, obtaining a perfectly precise model is very difficult due to modeling errors and external disturbances. Therefore, a crucial challenge when using MPC for system control is how to handle these disturbances.

When implementing MPC, it is essential to effectively manage external disturbances to ensure system robustness. Common robust MPC methodologies include tube-based MPC and constraint-tightening approaches, which enhance disturbance rejection capabilities by imposing specific constraints on system states and control inputs. Furthermore, disturbance rejection techniques can mitigate the impact of external disturbances. This can be achieved, for example, by designing a disturbance observer for disturbance compensation, or by constructing a disturbance model through learning the disturbance dynamic characteristics, thereby enabling disturbance prediction and suppression. In many chemical processes, disturbances exhibit relatively slow variations, making the development of efficient and practical methods to address such disturbance problems of significant engineering value.

This chapter introduces a disturbance-compensated composite event-triggered MPC algorithm specifically designed for discrete systems subject to slowly varying disturbances. The controller comprises two components: (i) a disturbance compensation controller using the slow-varying nature of the disturbances to estimate them and subsequently apply a compensatory action based on these estimations, and (ii) a model predictive controller responsible for obtaining the optimal control sequence by minimizing a defined cost function. To mitigate computational burden and circumvent the need for solving the optimization problem at every sampling instant, an event-triggered mechanism is employed to facilitate intermittent optimization. Furthermore, to counteract the potential invalidation of triggering conditions due to discontinuous measurements in discrete systems, this chapter utilizes a predictive event-triggered mechanism. This mechanism estimates the system state for the next time step, thereby enabling robust triggering.

4.1 Problem Statement

Consider a discrete-time linear system:

$$x(k+1) = Ax(k) + Bu(k) + n(k), \tag{4.1}$$

where $x(k) \in \mathbb{X}$, $u(k) \in \mathbb{U}$, and $n(k)$ represent the state vector, control input, and disturbance, respectively. The state constraint $\mathbb{X}$ and the control constraint $\mathbb{U}$ are closed sets, and the eigenvalues of matrix A lie outside the unit circle.

Assumption 4.1 The disturbance is bounded and varies slowly over time, i.e.,

$$n(k) \in W = \{n \mid \|n\| \leq \eta\}, \quad \text{and} \quad \|n(k+1) - n(k)\| \leq \mu,$$

where $\mu > 0, \eta > 0$ are constants.

For some chemical processes with slowly varying disturbances or physical systems with small sampling intervals, Assumption 4.1 is reasonable.

Assumption 4.2 The disturbance lies within the image space of B, that is, for all $n \in W$, there exists a control input $u \in H \subseteq \mathbb{U}$ such that $Bu = -n$.

In this chapter, the disturbance-compensating control set is defined by:

$$H = \{u \mid Bu = -n, \ \ n \in W\}.$$

The nominal model is applied by neglecting the disturbance:

$$\hat{x}(k+1) = A\hat{x}(k) + B\hat{u}(k). \tag{4.2}$$

Define the function $g(x, u) = (A - I)x + Bu$. It then follows that:

$$\hat{x}(k+1) = g(\hat{x}(k), \hat{u}(k)) + \hat{x}(k).$$

Assumption 4.3 The nominal system (4.2) is stabilizable, i.e., there exists a gain K such that $A + BK$ is Hurwitz. Moreover, there exists a set

$$\Omega_r = \{x \mid V_N(x) = \|x\|_P^2 \leq r^2\},$$

such that for all $x \in \Omega_r$, it holds that $Kx \in \mathbb{U} \ominus H$. If the control input $K\hat{x}(k)$ is applied to the nominal system (4.2), then the following condition is satisfied:

$$V_N(\hat{x}(k+1)) - V_N(\hat{x}(k)) \leq -\|\hat{x}(k)\|_{Q^*}^2,$$

where $Q^* = K^T RK + Q$.

Lemma 4.1 *If Assumption 4.3 holds, then for the system (4.1), under the control law $u(k) = Kx(k)$, if the following condition is satisfied:*

$$\|A + BK\|^2 r^2 + 2\eta \|A + BK\| r + \eta^2 \leq \frac{r^2}{\lambda_{\max}(P)} \tag{4.3}$$

then the set Ω_r is a robust invariant set.

Proof: Assuming $x(k) \in \Omega_r$, the state at the next time step satisfies:

$$\begin{aligned} \|x(k+1)\|_P^2 &\leq \lambda_{\max}(P)\|(A + BK)x(k) + n\|^2 \\ &\leq \lambda_{\max}(P)(\|A + BK\|^2 r^2 + 2\eta \|A + BK\| r + \eta^2). \end{aligned} \tag{4.4}$$

If condition (4.3) holds, then $\|x(k+1)\|_P^2 \leq r^2$, implying that $x(k+1) \in \Omega_r$, i.e., $x(k) \in \Omega_r$ indicates that $x(k+1) \in \Omega_r$. As a result, Ω_r is a robust invariant set. □

The objective of this chapter is to stabilize the system state to the neighborhood of the origin in the presence of external disturbances. A dual-mode MPC strategy is employed to achieve this goal: outside the robust invariant set, a disturbance-compensated composite event-triggered MPC scheme is used to minimize the frequency of optimization problem-solving as much as possible, while inside the invariant set, a state feedback controller is applied.

4.2 Composite Event-triggered MPC Mechanism

For the discrete-time system described in (4.1), this section designs a composite event-triggered MPC control law, which consists of two components: a disturbance compensation controller and a model predictive controller. A predictive triggering mechanism is introduced to realize event-triggered updates for the discrete system, thereby enabling the update of the control law and the solution of the associated optimization problem.

4.2.1 Disturbance Compensation Controller Design

The core idea of the disturbance compensation controller in this section is to estimate the disturbance and use a portion of the control input to compensate for the estimated disturbance.

According to Assumption 4.1, the disturbance varies slowly. Therefore, the disturbance at a previous time instant can be used as an estimate of the current disturbance, and it can be expressed by:

$$\tilde{n}(k) = x(k) - Ax(k-1) - Bu(k-1), \tag{4.5}$$

where $\tilde{n}(k)$ represents the disturbance estimated at time k.

Consider Assumptions 4.1 and 4.2. The disturbance compensation control input can be constructed by:

$$u^{\mathrm{com}}(k) = -B^{+}\tilde{n}(k), \tag{4.6}$$

where B^{+} denotes the Moore–Penrose inverse of matrix B.

Since the energy of disturbance is limited, and in light of Assumption 4.2, a portion of the control input can be used to compensate for the matched disturbance. For general disturbances, the compensatory control input can be constructed by $u = \arg\min \|Bu - n\|$ to achieve disturbance compensation.

4.2.2 Model Predictive Controller Design

Define $\{k_l\}$, $l \in \mathbb{N}$ as the sequence of triggering instants for the event-triggered MPC, with $k_0 = 0$. The optimization objective function is designed by:

$$J(\hat{x}(k_l), \hat{\mathbf{u}}(k_l)) = \sum_{i=0}^{N-1} L(\hat{x}(k_l+i|k_l), \hat{u}(k_l+i|k_l)) + V_N\left(\hat{x}\left(k_l+N|k_l\right)\right), \tag{4.7}$$

where $L(x,u) = \|x\|_Q^2 + \|u\|_R^2$ denotes the stage cost function, and $V_N(x) = \|x\|_P^2$ represents the terminal cost. The terms $\hat{x}(k_l+i|k_l)$ and $\hat{u}(k_l+i|k_l)$ denote the predicted state and control input at time k_l+i, respectively, based on information available at time k_l. The matrices P, Q, R are positive definite, and the prediction horizon is N.

At the triggering time k_l, the optimization problem can be formulated by:

$$\hat{\mathbf{u}}^*(k_l) = \arg\min_{\hat{\mathbf{u}}(k_l)} J(\hat{x}(k_l), \hat{\mathbf{u}}(k_l)), \tag{4.8}$$

$$\text{s.t.}\quad \hat{x}(k_l+i+1|k_l) = A\hat{x}(k_l+i|k_l) + B\hat{u}(k_l+i|k_l), \tag{4.9}$$

$$\hat{x}(k_l+i|k_l) \in \mathbb{X} \ominus X_i, \tag{4.10}$$

$$\hat{u}(k_l+i|k_l) \in \mathbb{U} \ominus H, \tag{4.11}$$

$$\hat{x}(k_l+N|k_l) \in \Omega_f, \tag{4.12}$$

where $\hat{x}(k_l\,|k_l) = x(k_l)$. The optimal control sequence obtained from solving the optimization problem is denoted by $\hat{\mathbf{u}}^*(k_l)$. The state contraction set is designed by:

$$X_i = \left\{ x \mid \|x\|_P \le \sqrt{\lambda_{\max}(P)}\,\mu \sum_{s=0}^{i-1} \left(\sqrt{\frac{\lambda_{\max}(P)}{\lambda_{\min}(P)}} \|A\| \right)^s (i-s) \right\},$$

which characterizes the impact of disturbances on the system state. The size of this contraction set is determined based on the assumption of slowly varying disturbances. Compared with methods that do not incorporate disturbance compensation, the contraction applied here is smaller, thereby yielding a larger admissible state region in the optimization problem. The set $\mathbb{U} - H$ represents the admissible control set in the optimization problem, obtained by subtracting the disturbance compensation set from the original control set. Finally, the terminal set is designed by:

$$\Omega_f = \left\{ x \mid \|x\|_P^2 \le \varepsilon^2,\ 0 < \varepsilon < r \right\}.$$

Due to the presence of disturbances, there exists a deviation between the actual disturbed state and the predicted state of the nominal system. The nominal state constraint set in the optimization problem (4.10), denoted by $\mathbb{X} - X_i$, is obtained by subtracting the state contraction set X_i from the original state constraint set $\mathbb{X}$.

Assumption 4.4 (Initial feasibility) The optimization problem (4.8)–(4.12) is feasible and admits a solution at the initial time t_0. Moreover, $\Omega_r \subseteq \mathbb{X} \ominus X_N$, where X_N denotes the state contraction set at $i = N$ as defined in (4.10).

4.2.3 Composite Event-triggered MPC

In the framework proposed in this chapter, a composite event-triggered controller is employed, where the control input $u(k)$ consists of two components. The first component, $u^{\text{com}}(k)$, is designed to compensate for partial disturbances, while the second component, $u^{\text{mpc}}(k)$, corresponds to the MPC input obtained by solving the optimization problem. Due to the use of an event-triggered mechanism, the overall control input applied between the current triggering instant k_l and the next triggering instant k_{l+1} is expressed as:

$$u(k) = u^{\text{com}}(k) + u^{\text{mpc}}(k), \quad k \in [k_l, k_{l+1}) \tag{4.13}$$

where over the interval $[k_l,\ k_{l+1})$, the disturbance compensation control input is given by $u^{\text{com}}(k) = -B^+\tilde{n}(k_l)$, and the MPC input is given by $u^{\text{mpc}}(k) = \hat{u}^*(k)$. The triggering interval is denoted by $m(k_l) = k_{l+1} - k_l$.

The disturbance compensation controller $u^{\text{com}}(k)$ is computed only at each triggering instant and remains constant between two consecutive triggering instants, i.e., it is implemented in an event-triggered manner, synchronized with the triggering interval of the MPC controller. If the rate of disturbance variation is sufficiently small and satisfies Assumption 4.1, the disturbance compensation strategy is effective. In addition, $u^{\mathrm{mpc}}(k)$ corresponds to the optimal control sequence obtained by solving the optimization problem, and the portion applied over the interval from k_l to k_{l+1} is used directly.

Due to the fact that the actual disturbed system employs the composite control input (4.13), while the nominal system uses the MPC input (4.8), the deviation between the actual state and the predicted nominal state is defined by:

$$e(k_l + i) = x(k_l + i) - \hat{x}(k_l + i \mid k_l).$$

The bound of this state deviation can be characterized using Lemma 4.2.

Lemma 4.2 *If the composite control law (4.13) is applied to the actual system (4.1), and the optimal control sequence* $\hat{\mathbf{u}}^*(x(k_l))$ *obtained from the optimization problem (4.8)–(4.12) is applied to the nominal system (4.2), then the deviation between the actual and predicted states,* $e(k_l + i)$*, satisfies:*

$$\|e(k_l + i)\|_P \leq \sqrt{\lambda_{\max}(P)}\mu \sum_{s=0}^{i-1} \left(\sqrt{\frac{\lambda_{\max}(P)}{\lambda_{\min}(P)}} \|A\| \right)^s (i - s). \tag{4.14}$$

Proof: At time $k_l + i$, the actual state under the composite event-triggered control input can be calculated by:

$$x(k_l + i) = Ax(k_l + i - 1) + Bu(k_l + i - 1) + n(k_l + i - 1) \tag{4.15}$$

where $u(k_l + i - 1) = u^{\mathrm{com}}(k_l + i - 1) + u^{\mathrm{mpc}}(k_l + i - 1)$.

At time $k_l + i$, the predicted state under the MPC input can be calculated by:

$$\hat{x}(k_l + i) = A\hat{x}(k_l + i - 1) + B\hat{u}^*(k_l + i - 1). \tag{4.16}$$

By taking the difference between (4.15) and (4.16), the deviation between the actual and predicted states can be shown to satisfy:

$$\|e(k_l + i)\|_P = \|Ae(k_l + i - 1) + Bu^{\mathrm{com}}(k_l + i - 1) + n(k_l + i - 1)\|_P. \tag{4.17}$$

Since the disturbance compensation controller applies a constant input between two triggering instants, it follows that $u^{com}(k_l + i - 1) = u^{com}(k_l)$. Moreover, as the controller uses the disturbance value from the previous time step as the estimate for the next, the state deviation in (4.17) can thus be calculated by:

$$\begin{aligned}\|e(k_l + i)\|_P &= \|Ae(k_l + i - 1) + Bu^{com}(k_l) + n(k_l + i - 1)\|_P \\ &= \|Ae(k_l + i - 1) + n(k_l + i - 1) - n(k_l - 1)\|_P.\end{aligned} \tag{4.18}$$

By incorporating the upper bound on the disturbance variation rate from Assumption 4.1 and applying the triangle inequality, the state deviation in (4.18) satisfies:

$$\|e(k_l + i)\|_P \leq \sqrt{\frac{\lambda_{\max}(P)}{\lambda_{\min}(P)}} \|A\| \, \|e(k_l + i - 1)\|_P + i\sqrt{\lambda_{\max}(P)}\mu. \tag{4.19}$$

Having established the relationship between $e(k_l + i)$ and $e(k_l + i - 1)$, the state deviation can be derived using a iterative approach to satisfy:

$$\begin{aligned}
&\|e(k_l + i)\|_P \\
&\leq \left(\sqrt{\frac{\lambda_{\max}(P)}{\lambda_{\min}(P)}} \|A\|\right)^2 \|e(k_l + i - 2)\|_P + \sqrt{\lambda_{\max}(P)}\mu \sum_{s=0}^{1} \left(\sqrt{\frac{\lambda_{\max}(P)}{\lambda_{\min}(P)}} \|A\|\right)^s (i - s) \\
&\leq \left(\sqrt{\frac{\lambda_{\max}(P)}{\lambda_{\min}(P)}} \|A\|\right)^3 \|e(k_l + i - 3)\|_P + \sqrt{\lambda_{\max}(P)}\mu \sum_{s=0}^{2} \left(\sqrt{\frac{\lambda_{\max}(P)}{\lambda_{\min}(P)}} \|A\|\right)^s (i - s) \\
&\cdots\cdots \\
&\leq \left(\sqrt{\frac{\lambda_{\max}(P)}{\lambda_{\min}(P)}} \|A\|\right)^i \|e(k_l)\|_P + \sqrt{\lambda_{\max}(P)}\mu \sum_{s=0}^{i-1} \left(\sqrt{\frac{\lambda_{\max}(P)}{\lambda_{\min}(P)}} \|A\|\right)^s (i - s).
\end{aligned} \tag{4.20}$$

Since $x(k_l) = \hat{x}(k_l)$ holds at time k_l, the state deviation (4.20) satisfies:

$$\|e(k_l + i)\|_P \leq \sqrt{\lambda_{\max}(P)}\mu \sum_{s=0}^{i-1} \left(\sqrt{\frac{\lambda_{\max}(P)}{\lambda_{\min}(P)}} \|A\|\right)^s (i - s), \tag{4.21}$$

which completes the proof. □

At the next triggering instant k_{l+1}, the disturbance compensation controller (4.6) is recalculated, and a new optimization problem (4.8)–(4.12) is solved. In the control framework proposed in this chapter, the composite event-triggered controller (4.13) is employed when the state lies outside the robust invariant set Ω_r. Once the state enters Ω_r, the linear feedback controller $u = Kx$ is adopted to reduce computational burden, thereby forming a dual-mode control mechanism.

4.2.4 Event-triggered Mechanism with Estimation

In the context of existing event-triggered frameworks for continuous-time systems, the design principle for triggering conditions is to treat the moment when the state deviation equals the triggering threshold as the triggering instant. However, since state deviations in discrete-time systems are inherently discontinuous, this principle is not applicable. Inspired by the one-step-ahead prediction approach in [52], this section proposes a novel event-triggered anticipation method tailored for composite MPC controllers. The core idea is to evaluate the current state deviation and estimate the next-step deviation in advance by considering the worst-case effect of disturbances. If the deviations at both the current and the next sampling instants are below the triggering threshold, the next sampling instant is safe, and no triggering occurs at the present time. Conversely, if the current deviation is below the threshold, but the predicted deviation at the next instant exceeds it, then the next sampling instant is considered unsafe, and the current instant is selected as the triggering time.

Suppose that at time step $k_l + i$, the deviation between the actual state and the predicted state is given by $e(k_l + i) = x(k_l + i) - \hat{x}(k_l + i \mid k_l)$. Then, at time step $k_l + i + 1$, the state deviation satisfies:

$$\begin{aligned}\|e(k_l + i + 1)\|_P &= \|Ae(k_l + i) + Bu(k_l + i) + n(k_l + i) - Bu^{\mathrm{mpc}}(k_l + i)\|_P \\ &\leq \|Ae(k_l + i)\|_P + \sqrt{\lambda_{\max}(P)}(i + 1)\mu.\end{aligned} \tag{4.22}$$

The triggering condition can be designed by:

$$k_{l+1} = \min\{\tilde{k}_{l+1}, k_l + N\}, \tag{4.23}$$

$$\tilde{k}_{l+1} = \left\{ k_l + i \,\middle|\, \begin{array}{l} \|e(k_l + i)\|_P \leq \sigma \\ \|Ae(k_l + i)\|_P + \sqrt{\lambda_{\max}(P)}(i + 1)\mu > \sigma \end{array} \right\}, \tag{4.24}$$

where the constant μ denotes the upper bound on the disturbance variation rate as specified in Assumption 4.1. The triggering threshold σ, designed based on feasibility considerations, is designed by:

$$\sigma = \frac{r - \varepsilon}{\left(1 + \|A - I\| \sqrt{\frac{\lambda_{\max}(P)}{\lambda_{\min}(P)}}\right)^N}. \tag{4.25}$$

To ensure that the triggering interval satisfies $m(k_l) \geq 1$, the triggering condition requires that $\|e(k_l + 1)\| P \leq \sigma$. Accordingly, based on (4.22), it must hold that $\sqrt{\lambda \max(P)}\, \mu \leq \sigma$.

Based on the above analysis, the complete procedure of the control algorithm is summarized in Algorithm 4.

Algorithm 4 Composite Event-triggered MPC Algorithm

1: At time step k_l, measure the system state $x(k_l)$. If $x(k_l) \in \Omega_r$, proceed to Step 4. If $x(k_l) \notin \Omega_r$, compute the disturbance-compensating controller according to (4.6), solve the MPC optimization problem based on (4.8)–(4.12), and determine the composite event-triggered controller using (4.13). Initialize $i = 0$.
2: Apply the composite controller $u(k_l + i)$ to the actual system (4.1). Check the triggering conditions (4.23)–(4.24).
3: If the triggering condition is satisfied, set $k_{l+1} = k_l + i$, update $l = l + 1$, and return to Step 1. Otherwise, set $i = i + 1$ and return to Step 2.
4: Apply the control law $u(k_l) = Kx(k_l)$ to the actual system. Set $k_{l+1} = k_l + 1$, update $l = l + 1$, and return to Step 1.

4.3 Feasibility and Stability

This section analyzes the feasibility of the optimization problem and the stability of the system under the proposed composite event-triggered mechanism.

4.3.1 Recursive Feasibility

Feasibility, in this context, refers to the property that if the optimization problem (4.8)–(4.12) is solvable at the initial time step, then it remains solvable at all subsequent time steps.

Lemma 4.3 *[74] Suppose the sequence $\{a_n\}, \{b_n\}, \{c_n\}, n \in N$ is nonnegative, and satisfies:*

$$a_n \leq b_n + \sum_{0 \leq j < n} c_j a_j.$$

If $b_n = b$ is constant, it holds that:

$$a_n \leq b \prod_{0 \leq i < n} (1 + c_i).$$

Lemma 4.4 *Suppose that $x_a \in \mathbb{X} \ominus X_{i+j}$, and*

$$\|x_a - x_b\|_P \leq \sqrt{\lambda_{\max}(P)} \sum_{s=0}^{j-1} \left(\sqrt{\frac{\lambda_{\max}(P)}{\lambda_{\min}(P)}} \|A\| \right)^{i+s} (j - s)\mu.$$

It then holds that $x_b \in \mathbb{X} \ominus X_i$.

Proof: Let $y = x_b - x_a + x_c$, where $x_c \in X_i$. Then, by the triangle inequality, it follows that:

$$\|y\|_P = \|x_b - x_a + x_c\|_P \leq \|x_a - x_b\|_P + \|x_c\|_P \,. \tag{4.26}$$

Given that $x_c \in X_i$, it follows that:

$$\|x_c\|_P \leq \sqrt{\lambda \max(P)}\, \mu \sum_{s=0}^{i-1} \left(\sqrt{\frac{\lambda_{\max}(P)}{\lambda_{\min}(P)}} \|A\| \right)^{s} (i - s),$$

It then follows that:

$$\|x_c\|_P \leq \sqrt{\lambda_{\max}(P)} \mu \sum_{s=0}^{i-1} \left(\sqrt{\frac{\lambda_{\max}(P)}{\lambda_{\min}(P)}} \|A\| \right)^{s} (i + j - s). \tag{4.27}$$

For $\|x_a - x_b\|_P$, by rearranging the summation order, it holds that:

$$\begin{aligned} \|x_a - x_b\|_P &\leq \sqrt{\lambda_{\max}(P)} \mu \sum_{s=0}^{j-1} \left(\sqrt{\frac{\lambda_{\max}(P)}{\lambda_{\min}(P)}} \|A\| \right)^{i+s} (j - s) \\ &\leq \sqrt{\lambda_{\max}(P)} \mu \sum_{s=i}^{i+j-1} \left(\sqrt{\frac{\lambda_{\max}(P)}{\lambda_{\min}(P)}} \|A\| \right)^{s} (i + j - s). \end{aligned} \tag{4.28}$$

Substituting (4.27) and (4.28) into (4.26) yields:

$$\|y\|_P \leq \sqrt{\lambda_{\max}(P)} \mu \sum_{s=0}^{i+j-1} \left(\sqrt{\frac{\lambda_{\max}(P)}{\lambda_{\min}(P)}} \|A\| \right)^{s} (i + j - s), \tag{4.29}$$

implying that $y \in X_{i+j}$. Given the condition $x_a \in \mathbb{X} \ominus X_{i+j}$, it follows that $y + x_a \in \mathbb{X}$.

Based on the definition $y = x_b - x_a + x_c$, it holds that $y + x_a = x_b + x_c$. Since $y + x_a \in \mathbb{X}$, it follows that $x_b + x_c \in \mathbb{X}$. Given that $x_c \in X_i$, it can be concluded that $x_b \in \mathbb{X} \ominus X_i$. □

The feasibility of the optimization problem can be further established by leveraging Lemma 4.3 and Lemma 4.4.

Theorem 4.1 *Suppose that Assumptions 4.1–4.4 are satisfied. If*

$$\varepsilon^2 \geq \left(1 - \frac{\lambda_{\min}(Q^*)}{\lambda_{\max}(P)}\right) r^2, \tag{4.30}$$

then the optimization problem (4.8)–(4.12) is feasible at all triggering instants.

Proof: The proof proceeds by mathematical induction. According to Assumption 4.4, the optimization problem is feasible at time k_0. Suppose that at time k_l, for $l \geq 0$, the optimization problem (4.8–4.12) is feasible. To ensure recursive feasibility, it should be proved that the optimization problem remains feasible at the next solution time k_{l+1}; i.e., the terminal constraint, state constraints, and control constraints must be satisfied at time k_{l+1}.

Based on the optimal control sequence and the state-feedback controller obtained at time k_l, a feasible control sequence for the optimization problem at time k_{l+1} can be constructed by:

$$\tilde{u}(k_{l+1}+i|k_{l+1}) = \begin{cases} \hat{u}^*(k_{l+1}+i|k_l), & i \in [0, N-m(k_l)-1] \\ K\tilde{x}(k_{l+1}+i|k_{l+1}), & i \in [N-m(k_l), N] \end{cases}, \tag{4.31}$$

where $\tilde{x}(k_{l+1}+i|k_{l+1})$ denotes predictive states.

First, consider the terminal constraint (4.12). For $k \in [k_{l+1}, k_l+N]$, it follows from Lemma 4.3 that:

$$\|\tilde{x}(k_{l+1}+i\,|k_{l+1}) - \hat{x}^*(k_{l+1}+i\,|k_l)\,\|_P \le \sigma\left(1+\|A-I\|\sqrt{\frac{\lambda_{\max}(P)}{\lambda_{\min}(P)}}\right)^i. \tag{4.32}$$

Substituting $i = N - m(k_l)$ into (4.32), and considering the triggering threshold σ, yields:

$$\|\tilde{x}(k_l+N\,|k_{l+1}\,)\|_P \le r - \varepsilon + r \le r, \tag{4.33}$$

indicating that $\tilde{x}(k_l+N\,|k_{l+1}\,) \in \Omega_r$.

Next, consider the interval $k \in [k_l+N+1, k_{l+1}+N]$. According to Assumption 4.3, it is evident that $\lambda_{\min}(Q^*) \le \lambda_{\max}(P)$. Substituting $k = k_l + N + i$ into Assumption 4.3 yields:

$$V_N(\tilde{x}(k_l+N+i+1\,|k_{l+1})) \le \left(1-\frac{\lambda_{\min}(Q^*)}{\lambda_{\max}(P)}\right) V_N(\tilde{x}(k_l+N+i\,|k_{l+1})). \tag{4.34}$$

By multiplying (4.34) from $i = m(k_l) - 1$ to $i = 0$, and applying conditions (4.30) and (4.33), we can obtain that:

$$V_N(\tilde{x}(N+k_{l+1}\,|k_{l+1})) \le \left(1-\frac{\lambda_{\min}(Q^*)}{\lambda_{\max}(P)}\right) r^2 \le \varepsilon^2, \tag{4.35}$$

indicating that $\tilde{x}(k_{l+1}+N\,|k_{l+1}\,) \in \Omega_\varepsilon$, thus satisfies the terminal constraint.

Next, we analyze the state constraints. For $i \in N_{[1,N-m(k_l)]}$, it follows from Lemma 4.2 that:

$$\begin{aligned} \|\tilde{x}(k_{l+1}+i\,|k_{l+1}) - \hat{x}^*(i+k_{l+1}\,|k_l\,)\|_P &\le \left(\sqrt{\frac{\lambda_{\max}(P)}{\lambda_{\min}(P)}}\|A\|\right)^i \|e(k_l+m(k_l))\|_P \\ &\le \sqrt{\lambda_{\max}(P)} \sum_{s=0}^{m(k_l)-1} \left(\sqrt{\frac{\lambda_{\max}(P)}{\lambda_{\min}(P)}}\|A\|\right)^{s+i} (m(k_l)-s)\mu. \end{aligned} \tag{4.36}$$

Since $\hat{x}^*(k_{l+1}+i\,|k_l)\in\mathbb{X}-X_{i+m(k_l)}$, it follows from Lemma 4.4 and (4.36) that $\tilde{x}(k_{l+1}+i\,|k_{l+1})\in X\ominus X_i$. For $i\in N_{[-m(k_l)+N,N]}$, Lemma 4.1 implies that under the admissible feedback control law $u=K\tilde{x}$, the predicted states always remain within the set Ω_r. According to Assumption 4.4, we have $\Omega_r\subseteq\mathbb{X}\ominus X_N$. Consequently, it follows that $\tilde{x}(i+k_{l+1}\,|k_{l+1})\in\mathbb{X}\ominus X_N\subseteq\mathbb{X}\ominus X_i$, implying that the state constraints are satisfied.

Finally, with regard to the control constraints, it follows directly from the construction procedure of the admissible control input in (4.31) that the control constraints are always satisfied.

In summary, with the implementation of the admissible control sequence (4.31), the terminal constraint, state constraints, and control constraints of the optimization problem are all satisfied at time k_{l+1}. Therefore, by the principle of mathematical induction, the optimization problem (4.8)–(4.12) is recursively feasible. □

4.3.2 Closed-loop Stability

The closed-loop stability is given by the following theorem.

Theorem 4.2 *If the composite event-triggered model predictive control given by Algorithm 4 is applied, and there exists a parameter $\gamma\in(0,+\infty)$ such that the following conditions are satisfied:*

$$\frac{\lambda_{\max}(Q)}{\lambda_{\min}(P)}\left[\frac{\sigma^2}{d^2-1}(d^{2N-2}-1)+\frac{2\sigma c}{d-1}(d^{N-1}-1)\right]+\sigma d^{N-1}(r+\varepsilon)<\frac{\lambda_{\min}(Q)}{\lambda_{\max}(P)}r^2-\gamma, \tag{4.37}$$

where

$$a=\sqrt{\frac{\lambda_{\min}(P)}{\lambda_{\max}(P)}}\|A\|,\quad b=\sup_{u\in\mathbb{U}}\|Bu\|_P,$$

$$c=\sup_{i\in N_{[1,N]}}\left\{\frac{1}{a^{N-i}}\left(\varepsilon+\frac{a^{N-i}-1}{a-1}b\right)\right\},\quad d=1+\|A-I\|\sqrt{\frac{\lambda_{\max}(P)}{\lambda_{\min}(P)}}.$$

Then the state of the actual disturbed system converges to the set Ω_r within a finite time. Furthermore, once the system state enters the set Ω_r, it remains within Ω_r under the state feedback controller.

Proof: Suppose that at the initial time k_0, the system state lies outside the set Ω_r, i.e., $x(k_0)\notin\Omega_r$. Use the cost function associated with the optimal control sequence $\hat{\mathbf{u}}^*(k_l)$ at time $x(k_l)$, denoted as $J(x(k_l),\hat{\mathbf{u}}^*(k_l))$, as the candidate Lyapunov function $V(x(k_l))$. The cost function at time $x(k_{l+1})$ corresponding to a feasible control sequence is expressed as $\tilde{V}(x(k_{l+1}))=J(x(k_{l+1}),\tilde{u}(k_{l+1}))$.

Based on the optimality of the cost function, it follows that $V(x(k_{l+1})) \leq \tilde{V}(x(k_{l+1}))$. Therefore, the deviation between $V(x(k_{l+1}))$ and $V(x(k_l))$ satisfies:

$$\begin{aligned} V(x(k_{l+1})) - V(x(k_l)) &\leq \tilde{V}(x(k_{l+1})) - V(x(k_l)) \\ &\leq \Delta V_1 + \Delta V_2 + \Delta V_3. \end{aligned} \tag{4.38}$$

where

$$\begin{aligned} \Delta V_1 &= -\sum_{i=0}^{m(k_l)-1} (\|\hat{u}^*(k_l+i|k_l|\|_R^2 + \|\hat{x}^*(k_l+i|k_l|\|_Q^2), \\ \Delta V_2 &= \sum_{i=m(k_l)}^{N-1} (-\|\hat{x}^*(k_l+i|k_l|\|_Q^2 + \|\tilde{x}(k_l+i|k_{l+1}|\|_Q^2), \\ \Delta V_3 &= \sum_{i=N}^{N+m(k_l)-1} (\|\tilde{u}(k_l+i|k_{l+1}|\|_R^2 + \|\tilde{x}(k_l+i|k_{l+1}|\|_Q^2) \\ &\quad - \|\hat{x}^*(k_l+N|k_l|\,\|_P^2 + \|\tilde{x}(k_{l+1}+N|k_{l+1}|\,\|_P^2. \end{aligned}$$

For ΔV_1, it can be calculated by:

$$\Delta V_1 \leq -\|\hat{x}^*(k_l\,|k_l)\|_Q^2 \leq -\frac{\lambda_{\min}(Q)}{\lambda_{\max}(P)} r^2. \tag{4.39}$$

Consider ΔV_2. Since $\|Ax\|_P \leq \|Ax + Bu\|_P + \|Bu\|_P$, the step-by-step iterative method can be employed to derive that the deviation between $\hat{x}^*(k_l + i \mid k_l)$ and $\hat{x}^*(k_l + N \mid k_l)$ satisfies:

$$\|\hat{x}^*(k_l+i\,|k_l)\|_P \leq \frac{1}{a^{N-i}}\left(\|\hat{x}^*(k_l+N\,|k_l)\|_P + \frac{a^{N-i}-1}{a-1}b\right), \tag{4.40}$$

where

$$a = \sqrt{\frac{\lambda_{\min}(P)}{\lambda_{\max}(P)}}\|A\|, \quad b = \sup_{u\in\mathbb{U}} \|Bu\|_P.$$

Given that $\|\hat{x}^*(k_l + N \mid k_l)\|_P \leq \varepsilon$, an upper bound for $\hat{x}^*(k_l + i \mid k_l)$ can be derived. Substituting this bound into the expression for ΔV_2, and in conjunction with (4.32), yields:

$$\begin{aligned} \Delta V_2 &\leq \sum_{i=m(k_l)}^{N-1} \frac{\lambda_{\max}(Q)}{\lambda_{\min}(P)} \sigma d^{i-m(k_l)}(\sigma d^{i-m(k_l)} + 2\,\|\hat{x}^*(k_l+i\,|k_l\,\|_P) \\ &\leq \frac{\lambda_{\max}(Q)}{\lambda_{\min}(P)} \sum_{i=0}^{N-m(k_l)-1} \sigma d^i(\sigma d^i + 2c) \\ &\leq \frac{\lambda_{\max}(Q)}{\lambda_{\min}(P)}\left[\frac{\sigma^2}{d^2-1}(d^{2N-2}-1) + \frac{2\sigma c}{d-1}(d^{N-1}-1)\right], \end{aligned} \tag{4.41}$$

where

$$c = \sup_{i \in \mathbb{N}_{[1,N]}} \left\{ \frac{1}{a^{N-i}} \left(\varepsilon + \frac{a^{N-i}-1}{a-1} b \right) \right\}, \quad d = 1 + \|A - I\| \sqrt{\frac{\lambda_{\max}(P)}{\lambda_{\min}(P)}}.$$

Finally, consider ΔV_3. Based on Assumption 4.3, applying cumulative summation yields:

$$\begin{aligned} &\sum_{i=N}^{m(k_l)+N-1} (\|\tilde{u}(k_l + i\,|k_{l+1})\|_R^2 + \|\tilde{x}(k_l + i\,|k_{l+1})\|_Q^2) \\ &\quad \leq -\|\tilde{x}(k_{l+1} + N\,|k_{l+1})\|_P^2 + \|\tilde{x}(k_l + N\,|k_{l+1})\|_P^2. \end{aligned} \tag{4.42}$$

Substituting (4.42) into ΔV_3 yields:

$$\Delta V_3 \leq \sigma d^{N-m(k_l)}(r + \varepsilon) \leq \sigma d^{N-1}(r + \varepsilon). \tag{4.43}$$

By substituting (4.39), (4.41), and (4.43) into (4.38), and incorporating condition (4.37), it follows that:

$$V(x(k_{l+1})) - V(x(k_l)) < -\gamma. \tag{4.44}$$

Applying the cumulative summation to (4.44) yields:

$$V(x(k_{l+1})) < V(x(k_0)) - (l + 1)\gamma. \tag{4.45}$$

If $l \to \infty$, then the right-hand side of (4.45) tends toward $-\infty$. However, by definition, the left-hand side of (4.45) is nonnegative. Hence, by contradiction, l must be bounded. Therefore, the system state enters the set Ω_r within a finite time.

Once the system state enters the set Ω_r, it follows from Lemma 4.1 that Ω_r is a robust invariant set. Therefore, under state feedback control, the system state remains within the terminal set Ω_r. □

4.4 A Simulation Example

In the field of chemical engineering, the physical process of a stirred tank reactor can be represented by a linear state-space equation, and its corresponding discrete error model can be expressed by:

$$\begin{bmatrix} x_1(k+1) \\ x_2(k+1) \end{bmatrix} = A \begin{bmatrix} x_1(k) \\ x_2(k) \end{bmatrix} + B \begin{bmatrix} u_1(k) \\ u_2(k) \end{bmatrix} + \begin{bmatrix} w_1(k) \\ w_2(k) \end{bmatrix}, \tag{4.46}$$

where

$$A = \begin{bmatrix} 0.9986 & -0.0001 \\ -0.0019 & 1.0074 \end{bmatrix}, B = \begin{bmatrix} -0.0043 & 0.0012 \\ 0.004 & 0.0223 \end{bmatrix}.$$

The states x_1 and x_2 represent the errors of the reactant concentration and the reaction temperature, respectively. The input variables u_1 and u_2 correspond to the coolant temperature and the concentration of the added reagent, respectively. The control inputs are constrained by $\|u_1\| \leq 4$ K and $\|u_2\| \leq 6$ kmol/m^3. The external disturbance $w(t) = [w_1(k); w_2(k)]$ satisfies the constraints $\|w(k)\| \leq 0.005$ and $\|w(k+1) - w(k)\| \leq 0.0005$, and in the simulation, it is generated using random values that reflect slow-varying characteristics. Additionally, the state constraints are $\|x_1\| \leq 2$ kmol/m^3 and $\|x_2\| \leq 3$ K. The weighting matrices are $Q = \text{diag}\{10, 5\}$ and $R = \text{diag}\{0.002, 0.001\}$, and the terminal weighting matrix is designed as:

$$P = \begin{bmatrix} 0.79 & 0.13 \\ 0.13 & 0.15 \end{bmatrix}.$$

The prediction horizon N is set to 20, and the terminal parameter r is 0.03. The initial state is given by $x_1 = 0.2$ kmol/m^3, $x_2 = 0.6$ K.

The system is controlled using the method described in Algorithm 4, and the state and control trajectories are illustrated in Figure 4.1. Analysis of the figure

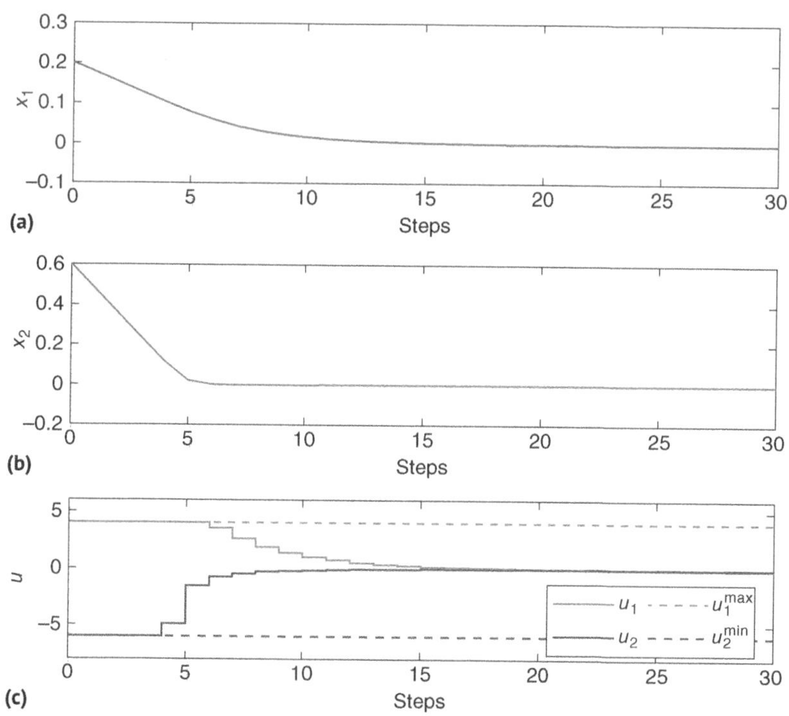

Figure 4.1 States and control inputs.

shows that the system state converges toward the origin, while the state and control constraints are consistently satisfied throughout the control process.

Different weight matrices influence the evolution of the state and control inputs. The performance of the system under two different sets of weights is compared and presented in Figure 4.2. The first set of weight matrices is designed as $Q_1 = \text{diag}\{10, 5\}$, $R_1 = \text{diag}\{0.002, 0.001\}$, and the terminal weight matrix:

$$P_1 = \begin{bmatrix} 0.79 & 0.13 \\ 0.13 & 0.15 \end{bmatrix}.$$

The second set is $Q_2 = \text{diag}\{5, 2.5\}$, $R_2 = \text{diag}\{0.02, 0.01\}$, with

$$P_2 = \begin{bmatrix} 0.89 & 0.11 \\ 0.23 & 0.16 \end{bmatrix}.$$

As shown in Figure 4.2, decreasing the state weights results in slower convergence of the state variables. Conversely, increasing the control weights allows the control inputs to converge quickly during the initial phase and rapidly move away from their boundary limits.

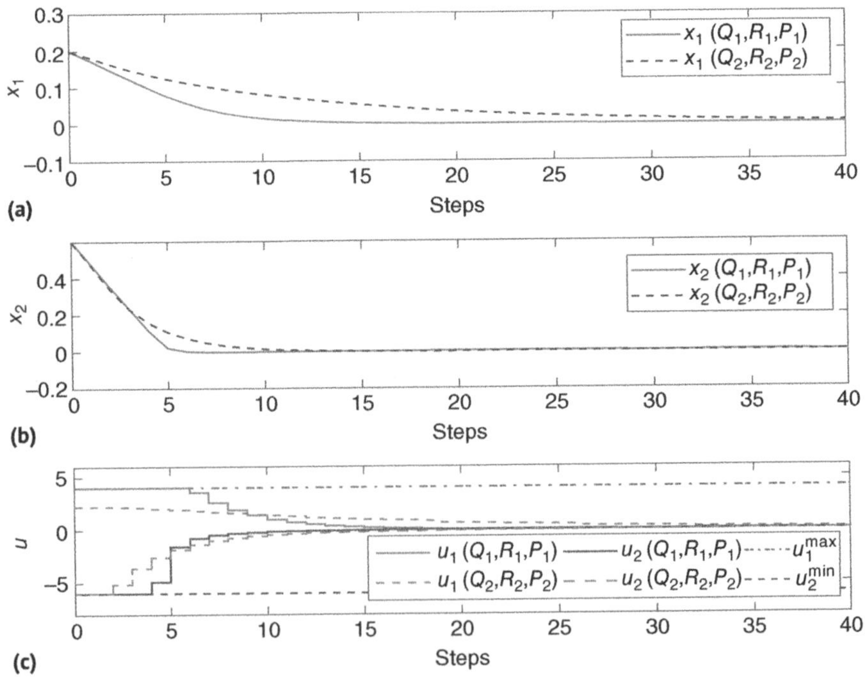

Figure 4.2 States and control inputs under different weight coefficients.

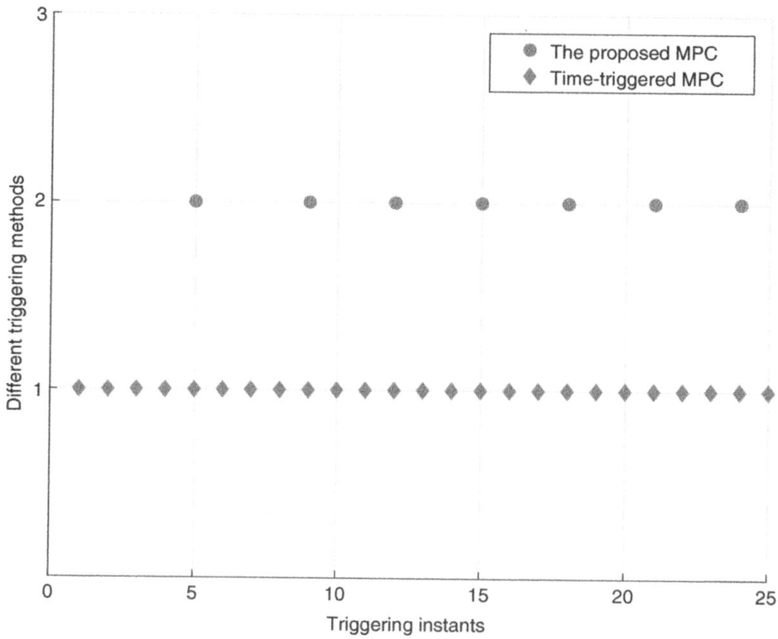

Figure 4.3 Triggering instants under the proposed composite event-triggered MPC and time-triggered MPC.

Traditional MPC approaches solve an optimization problem at every time step and are thus referred to as time-triggered MPC methods. The triggering instants of the proposed algorithm and those of the conventional time-triggered MPC are plotted in Figure 4.3. It is evident from the figure that the proposed composite event-triggered MPC solves only seven times of optimization, compared to 23 under the time-triggered MPC. This comparison highlights a significant reduction in the number of optimizations required by the proposed method. Since sampling instants impose minimal computational demand, whereas solving constrained optimization problems requires substantial processing power, the total computational time and complexity of the proposed method is approximately one-third that of the time-triggered MPC. Consequently, the proposed approach reduces computational load and conserves processing resources.

4.5 Conclusion

For discrete-time systems with constraints and slowly varying disturbances, this chapter proposes a composite event-triggered MPC framework. The controller consists of two components: a disturbance compensation controller, which estimates and compensates for external disturbances, and an MPC component, which solves optimization problems for the nominal system to achieve optimal control performance. The proposed algorithm leverages an event-triggering mechanism to reduce the frequency of optimization, thereby alleviating the computational burden on the processor.

To address the challenges of discretization, a predictive event-triggering strategy is adopted. By anticipating future behavior, this strategy enables timely triggering decisions and avoids issues caused by measurement discontinuities in discrete systems. Theoretical analysis is presented to establish the feasibility and stability of the closed-loop system under the proposed composite event-triggered MPC scheme. Simulation results further validate the effectiveness of the algorithm.

5

Event-triggered MPC with Periodic Sampling for Multi-agent Systems

Formation control tasks for multi-agent systems have broad applications in areas such as search, measurement, and exploration, offering significantly higher task complexity handling and execution efficiency compared to single-agent control systems. As a result, they possess considerable engineering value and scientific significance. Although Chapters 2–4 provide an in-depth study of event-triggered model predictive control (MPC) methods for individual systems, directly extending these methods to multi-agent systems introduces new challenges. This is because the triggering instants in individual event-triggered mechanisms are asynchronous. Applying such mechanisms to multi-agent scenarios leads to highly complex inter-agent communication and substantially increased sensor power consumption due to the need for real-time sensing and neighbor information reception. Therefore, it is necessary to design a distributed MPC strategy featuring a streamlined and efficient event-triggered communication mechanism that is tailored to the information exchange characteristics among agents.

This chapter investigates the formation control problem of a group of nonholonomic mobile robots using a distributed MPC (DMPC) approach. Within the distributed framework, a virtual central vehicle is constructed and designated as a virtual leader. Each actual vehicle tracks its desired relative position with respect to this virtual leader. Furthermore, a formation cost function is incorporated into the objective function of each vehicle, and coordinated formation motion with a fixed geometric configuration is achieved through inter-vehicle information exchange.

To alleviate the excessive computational burden on processors caused by continuous sampling and frequent optimization in classical MPC methods, this chapter adopts an event-triggered mechanism to reduce the frequency of solving the optimization problem. Under this mechanism, all vehicles operate on a unified temporal framework, where state sampling and information exchange are carried out at fixed periodic intervals. In traditional MPC, near-continuous sampling is typically approximated using very short and fixed intervals—often on the order of milliseconds or even microseconds. In contrast, the periodic sampling intervals proposed in this chapter can be flexibly designed according to system requirements and are significantly longer than those used in conventional methods. Furthermore, the theoretical soundness of this periodic sampling approach is rigorously established. This periodic sampling mechanism not only enables efficient information exchange among agents but also reduces hardware demands on sensors and processors through its intermittent operation, thereby lowering energy consumption. To address the discontinuity in state measurements inherent in periodic sampling, this chapter introduces a predictive estimation-based design for the triggering mechanism to ensure triggering safety. Finally, rigorous theoretical analysis is conducted for the proposed periodic sampled-data event-triggered model predictive control scheme, establishing the feasibility of the optimization problem and the stability of the closed-loop system for the mobile robots.

5.1 Problem Statement

This section focuses on a group of nonholonomic mobile robots sharing identical dynamics, where the dynamics of each robot i can be described by:

$$\dot{z}_i(t) = f_{\text{model}}(z_i(t), u_i(t)) = \begin{bmatrix} v_i(t)\cos\theta_i(t) \\ v_i(t)\sin\theta_i(t) \\ \omega_i(t) \end{bmatrix}, \tag{5.1}$$

where t denotes time, and $z_i(t) = [p_i(t); \theta_i(t)]$ and $u_i(t) = [v_i(t); \omega_i(t)]$ represent the system state and control input of vehicle i, respectively; $p_i(t) = [x_i(t); y_i(t)]$ denotes the position of the vehicle center in the global coordinate frame, while $\theta_i(t)$ indicates its heading angle; $v_i(t)$ and $\omega_i(t)$ correspond to the linear and angular velocities of vehicle i, respectively.

The control input constraints are defined as:

$$u_i(t) \in U_i = \{u_i(t) \mid |v_i(t)| \leq v_{\max},\ |\omega_i(t)| \leq \omega_{\max}\},$$

where $v_{\max}$ and $\omega_{\max}$ represent the maximum allowable linear and angular velocities, respectively. In certain existing models, the linear and angular velocities are coupled and confined within a diamond-shaped admissible region. For controller design, however, this work decouples the linear and angular velocities, effectively selecting a rectangular subset within the diamond-shaped constraint set, thereby ensuring independence between the control inputs. The communication topology among the vehicles is assumed to be fully connected, implying each vehicle i can exchange information bidirectionally with its neighbors j. The set of all neighbors of vehicle i is denoted by Ψ_i.

In order to accomplish the multi-agent tracking objective, this chapter adopts the leader–follower framework by introducing a virtual leader vehicle. The dynamics of this virtual leader are described by the following model:

$$\dot{z}_r(t) = f_{\text{model}}\left(z_r(t), u_r(t)\right) = \begin{bmatrix} \cos\theta_r(t) & 0 \\ \sin\theta_r(t) & 0 \\ 0 & 1 \end{bmatrix} \begin{bmatrix} v_r(t) \\ \omega_r(t) \end{bmatrix}, \tag{5.2}$$

where $z_r(t) = [p_r(t), \theta_r(t)]^T$ denotes the state of the virtual leader, where $p_r(t) = [x_r(t), y_r(t)]^T$ represents its position. The reference control input $u_r(t) = [v_r(t), \omega_r(t)]^T$ is predefined and transmitted to each follower vehicle in advance.

For each follower vehicle, the head position is defined as the point located a distance l from the vehicle center along the perpendicular bisector of the wheel axis. The corresponding state of the head position is given by:

$$z_{hi}(t) = \begin{bmatrix} p_{hi}(t) \\ \theta_{hi}(t) \end{bmatrix} = z_i(t) + l \begin{bmatrix} \cos\theta_i(t) \\ \sin\theta_i(t) \\ 0 \end{bmatrix}.$$

Let the head position of each vehicle be expressed by:

$$p_{hi}(t) = \begin{bmatrix} x_{hi}(t) \\ y_{hi}(t) \end{bmatrix};$$

the kinematic equation governing the head position can then be formulated as:

$$\dot{z}_{hi}(t) = \begin{bmatrix} \cos\theta_i(t) & -l\sin\theta_i(t) \\ \sin\theta_i(t) & l\cos\theta_i(t) \\ 0 & 1 \end{bmatrix} \begin{bmatrix} v_i(t) \\ \omega_i(t) \end{bmatrix}. \tag{5.3}$$

To establish the Frenet–Serret frames associated with the virtual leader and the followers, two moving coordinate systems, denoted as ${}^{r}O$ and ${}^{i}O$, are constructed. These coordinate frames are fixed at the center of the virtual leader and each follower vehicle, respectively, as illustrated in Figure 5.1. For each follower vehicle, a virtual reference point $p_{di} = [x_{di}, y_{di}]^T$ is defined within the coordinate frame

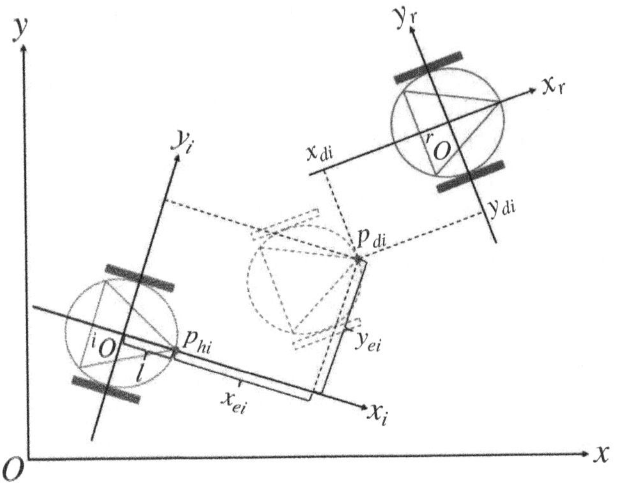

Figure 5.1 Positions of the virtual leader and follower i.

^rO of the virtual leader. The control objective for each vehicle is to guide its head position p_{hi} to the corresponding desired point p_{di}.

For each follower vehicle, the tracking error

$$p_{ei} = \begin{bmatrix} x_{ei} \\ y_{ei} \end{bmatrix}$$

is defined as the deviation between the head position $p_{hi}(t)$ of the follower and the corresponding virtual reference point p_{di}. This error is defined in the moving coordinate frame ^iO attached to each vehicle. The tracking error p_{ei} can be expressed as:

$$p_{ei}(t) = \begin{bmatrix} \cos\theta_i(t) & \sin\theta_i(t) \\ -\sin\theta_i(t) & \cos\theta_i(t) \end{bmatrix} (p_r(t) - p_{hi}(t)) + \begin{bmatrix} \cos\theta_{ei}(t) & -\sin\theta_{ei}(t) \\ \sin\theta_{ei}(t) & \cos\theta_{ei}(t) \end{bmatrix} p_{di}, \tag{5.4}$$

where $\theta_{ei}(t) = \theta_r(t) - \theta_i(t)$. Differentiating the tracking error yields:

$$\begin{aligned} \dot{p}_{ei}(t) = & \begin{bmatrix} 0 & \omega_i(t) \\ -\omega_i(t) & 0 \end{bmatrix} p_{ei}(t) \\ & + \begin{bmatrix} -v_i(t) + (v_r - y_{di}\omega_r)\cos\theta_{ei}(t) - x_{di}\omega_r \sin\theta_{ei}(t) \\ -l\omega_i(t) + (v_r - y_{di}\omega_r)\sin\theta_{ei}(t) + x_{di}\omega_r \cos\theta_{ei}(t) \end{bmatrix}. \end{aligned} \tag{5.5}$$

To facilitate interpretation and practical application, based on the derivative expression of the tracking error (5.5), the generalized input for each vehicle can be defined as:

$$u_{ei}(t) = f_{\text{gen}}(u_i(t)) = \begin{bmatrix} -v_i(t) + (v_r - y_{di}\omega_r)\cos\theta_{ei}(t) - x_{di}\omega_r \sin\theta_{ei}(t) \\ -l\omega_i(t) + (v_r - y_{di}\omega_r)\sin\theta_{ei}(t) + x_{di}\omega_r \cos\theta_{ei}(t) \end{bmatrix}, \tag{5.6}$$

which is actually the function of control inputs $v_i(t)$ and $\omega_i(t)$.

Taking into account that real-world vehicle motion is subject to external disturbances $n(t)$, it is assumed that such disturbances arise primarily from vehicle sideslip. Accordingly, only the translational components of the disturbance are considered, while those caused by roll motion are neglected. Under this assumption, the tracking error system can be represented by:

$$\dot{p}_{ei}(t) = \begin{bmatrix} 0 & \omega_i(t) \\ -\omega_i(t) & 0 \end{bmatrix} p_{ei}(t) + u_{ei}(t) + n_i(t), \tag{5.7}$$

where the disturbance $n_i(t) = [n_{xi}(t); n_{yi}(t)]$ is bounded, and it satisfies $\|n_i(t)\| \leq \eta$.

Due to the presence of disturbances, a robust DMPC framework is required to ensure that the tracking error $p_{ei}(t)$ of each vehicle converges to the neighborhood of the origin. Since each vehicle cannot predict future disturbances in advance, a nominal system framework must be constructed to facilitate the analysis of the distributed MPC problem. Denoting nominal states and inputs with a superscript $\hat{\cdot}$, the tracking error dynamics of each nominal system can be formulated as follows:

$$\dot{\hat{p}}_{ei}(t) = f(\hat{p}_{ei}(t), \hat{u}_i(t)) = \begin{bmatrix} 0 & \hat{\omega}_i(t) \\ -\hat{\omega}_i(t) & 0 \end{bmatrix} \hat{p}_{ei}(t) + \hat{u}_{ei}(t), \tag{5.8}$$

where $\hat{p}_{ei}(t)$ denotes the nominal tracking error of vehicle i, while $\hat{u}_{ei}(t)$ represents its nominal generalized input, which is given by:

$$\hat{u}_{ei}(t) = \begin{bmatrix} -\hat{v}_i(t) + (v_r - y_{di}\omega_r)\cos\hat{\theta}_{ei}(t) - x_{di}\omega_r \sin\hat{\theta}_{ei}(t) \\ -l\hat{\omega}_i(t) + (v_r - y_{di}\omega_r)\sin\hat{\theta}_{ei}(t) + x_{di}\omega_r \cos\hat{\theta}_{ei}(t) \end{bmatrix} \tag{5.9}$$

Assumption 5.1 The nominal tracking error system of each vehicle is Lipschitz continuous with respect to $\hat{p}_{ei}$, with a Lipschitz constant a. That is, for any two nominal error states $\hat{p}_{ei}^1$ and $\hat{p}_{ei}^2$, the following inequality holds:

$$\left\| f(\hat{p}_{ei}^1, \hat{u}_i) - f(\hat{p}_{ei}^2, \hat{u}_i) \right\| \leq a \left\| \hat{p}_{ei}^1 - \hat{p}_{ei}^2 \right\|.$$

At time t_k^i, an optimization problem is formulated for each vehicle i, with the objective of minimizing the following cost function:

$$J_i(\hat{p}_{ei}(t_k^i), \hat{u}_i(t_k^i)) = \int_{t_k^i}^{t_k^i+T} L_i(\hat{p}_{ei}(t|\, t_k^i), \hat{u}_i(t|\, t_k^i))dt + g_i(\hat{p}_{ei}(t_k^i + T|\, t_k^i)), \tag{5.10}$$

where the control horizon for the optimization problem is assigned by T, and the terminal cost is designed by:

$$g_i(\hat{p}_{ei}(t_k^i + T|\, t_k^i)) = \frac{1}{2}\left\|\hat{p}_{ei}(t_k^i + T|\, t_k^i)\right\|^2. \tag{5.11}$$

The stage cost is designed by:

$$L_i(\hat{p}_{ei}(t|\, t_k^i), \hat{u}_i(t|\, t_k^i)) = \left\|\hat{p}_{ei}(t|\, t_k^i)\right\|_Q^2 + \left\|\hat{u}_i(t|\, t_k^i)\right\|_P^2 + \sum_{j\in\Psi_i} \left\|\hat{p}_{ij}(t|\, t_k^i)\right\|_H^2, \tag{5.12}$$

where weighting matrices are defined by $Q = \text{diag}\{q_1, q_2\}$, $P = \text{diag}\{p_1, p_2\}$, and $H = \text{diag}\{h_1, h_2\}$. The term $\left\|\hat{p}_{ei}(t \mid t_k^i)\right\|_Q^2$ represents the state tracking error cost, with a smaller value indicates that the vehicle position is closer to the desired virtual point. Similarly, $\left\|\hat{u}_i(t \mid t_k^i)\right\|_P^2$ denotes the control effort cost, where a smaller value implies reduced control input magnitude. $\left\|\hat{p}_{ij}(t|\, t_k^i)\right\|_H^2$ is the formation error cost, which is defined as:

$$\left\|\hat{p}_{ij}(t|\, t_k^i)\right\|_H^2 = \left\|\hat{p}_{ei}(t|\, t_k^i) - \bar{p}_{ej}(t|\, t_k^i)\right\|_H^2, \tag{5.13}$$

where $\bar{p}_{ej}(t \mid t_k^i)$ denotes the assumed tracking error of vehicle j. At time t, if the position error vector of vehicle i is identical to that of vehicle j, the formation error cost becomes zero. In this case, vehicles i and j maintain a fixed relative formation, thereby achieving formation-preserving cooperative motion.

5.2 Distributed MPC Design

In this section, distributed MPC is designed for the nonholonomic multi-vehicle system. Due to the nonholonomic constraints, linearization around the origin is impossible for the vehicles, leading to distributed nonlinear MPC design for the vehicles.

5.2.1 Terminal Set and Auxiliary Terminal Control Design

At the beginning, we need to define the terminal set and the auxiliary terminal control.

Definition 5.1 The set region Ω is a terminal set with an auxiliary terminal control $\mu_i(\cdot)$, provided that if the nominal tracking error satisfies $\hat{p}_{ei}(\tau) \in \Omega$ at some time τ, then for any subsequent time t, the control $\hat{u}_i(t) = \mu_i(\hat{p}_{ei}(t))$ guarantees that the following conditions hold:

$$\hat{p}_{ei}(t) \in \Omega, \ \hat{u}_i(t) \in U_i. \tag{5.14}$$

For the nonholonomic vehicle, Lemma 5.1 provides the design of the terminal set as well as the corresponding auxiliary terminal controller.

Lemma 5.1 *For the nominal error dynamics (5.8), let*

$$A = \begin{bmatrix} 1 & 0 \\ -2y_{di} & x_{di}^2 + y_{di}^2 \end{bmatrix},$$

and choose the auxiliary terminal control:

$$u_i(t) = \mu_i(\hat{p}_{ei}(t \,|t_k^i)) = \begin{bmatrix} k_1\hat{x}_{ei} + (v_r - y_{di}\omega_r)\cos\hat{\theta}_{ei} - x_{di}\omega_r \sin\hat{\theta}_{ei} \\ \frac{1}{l}(k_2\hat{y}_{ei} + (v_r - y_{di}\omega_r)\sin\hat{\theta}_{ei} - x_{di}\omega_r \cos\hat{\theta}_{ei}) \end{bmatrix} \tag{5.15}$$

where $k_1, k_2 > 0$. Then, the terminal set can be calculated by:

$$\Omega = \left\{ (\hat{x}_{ei}, \hat{y}_{ei}) \left| \begin{array}{l} \frac{-v_{\max} - \sqrt{\|A\|}\min\|u_r\|}{k_1} \le \hat{x}_{ei} \le \frac{v_{\max} - \sqrt{\|A\|}\max\|u_r\|}{k_1}, \\ \frac{-l\omega_{\max} - \sqrt{\|A\|}\min\|u_r\|}{k_2} \le \hat{y}_{ei} \le \frac{l\omega_{\max} - \sqrt{\|A\|}\max\|u_r\|}{k_2}. \end{array} \right. \right\} \tag{5.16}$$

Proof: Taking the terminal cost as a Lyapunov function yields that:

$$g_i(\hat{p}_{ei}(t|\, t_k^i)) = \frac{1}{2}(\hat{x}_{ei}^2 + \hat{y}_{ei}^2). \tag{5.17}$$

Its time derivative can be calculated by:

$$\dot{g}_i(\hat{p}_{ei}(t|\, t_k^i)) = \hat{x}_{ei}\dot{\hat{x}}_{ei} + \hat{y}_{ei}\dot{\hat{y}}_{ei}. \tag{5.18}$$

By applying the auxiliary terminal controller (5.15) and substituting it into (5.9), the resulting nominal generalized input can be calculated by:

$$\hat{u}_{ei}(t) = [-k_1\hat{x}_{ei}; -k_2\hat{y}_{ei}]. \tag{5.19}$$

Substituting (5.8) and (5.9), which describe the nominal tracking error dynamics, into (5.19) yields:

$$\begin{aligned}\dot{g}_i(\hat{p}_{ei}(t|\, t_k^i)) &= \hat{x}_{ei}(\omega_i \hat{y}_{ei} - k_1 \hat{x}_{ei}) + \hat{y}_{ei}(-\omega_i \hat{x}_{ei} - k_2 \hat{y}_{ei}) \\ &= -k_1 \hat{x}_{ei}^2 - k_2 \hat{y}_{ei}^2 \leq 0,\end{aligned} \tag{5.20}$$

implying that, if $\hat{p}_{ei}(\tau) \in \Omega$, then for all $t \geq \tau$, it holds that $\hat{p}_{ei}(t) \in \Omega$.

The design rationale behind the terminal controller is to eliminate the nonlinear components in the generalized control input, thereby enabling it to be expressed in a linear form. Subsequently, it is necessary to demonstrate that when the terminal region satisfies the condition (5.16), the auxiliary terminal controller remains within the prescribed control constraint bounds.

Due to the boundedness of the control input, the auxiliary terminal control input must satisfy the constraint $\hat{u}_i(t) \in U_i$. Substituting the terminal controller and considering the upper and lower bounds of the control input yield:

$$-v_{\max} \leq k_1 \hat{x}_{ei} + (v_r - y_{di}\omega_r)\cos\hat{\theta}_{ei} - x_{di}\omega_r \sin\hat{\theta}_{ei} \leq v_{\max}, \tag{5.21}$$

$$-\omega_{\max} \leq \frac{1}{l}(k_2 \hat{y}_{ei} + (v_r - y_{di}\omega_r)\sin\hat{\theta}_{ei} - x_{di}\omega_r \cos\hat{\theta}_{ei}) \leq \omega_{\max}. \tag{5.22}$$

Applying the triangle inequality to (5.21) yields:

$$-v_{\max} \leq k_1 \hat{x}_{ei} + \sqrt{(v_r - y_{di}\omega_r)^2 + (x_{di}\omega_r)^2}\,\sin(\hat{\theta}_{ei} + \varphi) \leq v_{\max}. \tag{5.23}$$

By rearranging (5.23), it follows that $\hat{x}_{ei}$ satisfies:

$$\frac{-v_{\max} - \sqrt{\|A\|}\min\|u_r\|}{k_1} \leq \hat{x}_{ei} \leq \frac{v_{\max} - \sqrt{\|A\|}\max\|u_r\|}{k_1}. \tag{5.24}$$

Similarly, $\hat{y}_{ei}$ satisfies:

$$\frac{-l\omega_{\max} - \sqrt{\|A\|}\min\|u_r\|}{k_2} \leq \hat{y}_{ei} \leq \frac{l\omega_{\max} - \sqrt{\|A\|}\max\|u_r\|}{k_2}. \tag{5.25}$$

Based on (5.24) and (5.25), and in conjunction with condition (5.16), it can be concluded that when the terminal region satisfies condition (5.16), the terminal controller (5.15) ensures the satisfaction of control constraints. □

In fact, when the nominal state lies within the terminal region Ω, the auxiliary terminal controller $\mu_i(\hat{p}_{ei}(t))$ ensures, according to (5.20), that the nominal system state not only remains within the set Ω but also converges over time.

5.2.2 Distributed MPC Framework

In designing distributed MPC, each vehicle i cannot access the actual predicted trajectory of its neighboring vehicles. As a substitute, each neighbor transmits to vehicle i an assumed predicted trajectory, thereby enabling vehicle i to solve its optimization problem. To distinguish between variables, let $\hat{u}_i^*(t \mid t_k^i)$ and $\bar{u}_j(t \mid t_k^i)$ denote the optimal control input and the assumed input of neighbor j, respectively. Specifically, $\hat{u}_i^*(t \mid t_k^i)$ is obtained by solving the optimization problem, and its portion over the interval $[t_k^i, t_{k+1}^i)$ is applied as the actual control input to vehicle i.

The assumed control input trajectory $\bar{u}_j(t \mid t_k^i)$ transmitted by neighbor j to vehicle i is constructed as a combination of the neighbor's most recently computed optimal control input and terminal control input. The assumed control input $\bar{u}_j(t \mid t_k^i)$ is calculated by:

$$\bar{u}_j(t|\, t_k^i) = \begin{cases} \hat{u}_j^*(t|\, t_k^{ij}), & t \in [t_k^i, t_k^{ij} + T), \\ \mu_j(\bar{p}_{ej}(t \,|t_k^{ij})), & t \in [t_k^{ij} + T, t_k^i + T), \end{cases} \tag{5.26}$$

where t_k^{ij} denotes the most recent time prior to t_k^i at which neighboring vehicle j solved its optimization problem.

By employing the assumed control input trajectory, the corresponding assumed position error trajectory $\bar{p}_{ej}(t \mid t_k^i)$ can be derived by:

$$\bar{p}_{ej}(t|\, t_k^i) = \begin{cases} p_{ej}^*(t|\, t_k^{ij}), & t \in [t_k^i, t_k^{ij} + T), \\ p_{ej}^{k*}(t|\, t_k^{ij}), & t \in [t_k^{ij} + T, t_k^i + T), \end{cases} \tag{5.27}$$

where $p_{ej}^{k*}(t \mid t_k^{ij})$ denotes the position error trajectory corresponding to the terminal control input over the interval $[t_k^{ij} + T,\ t_k^i + T)$.

To ensure that the tracking error of vehicle i converges to a neighborhood of the origin, this section proposes a DMPC framework. The control signal for vehicle i is obtained by solving an optimization problem based on its nominal system, and a portion of the resulting optimal control input is applied to the actual system. At time t_k^i, the optimization problem for each vehicle i is defined as follows:

$$\min \quad J_i(\hat{p}_{ei}(t_k^i), \hat{u}_i(t_k^i)), \quad t \in [t_k^i, t_k^i + T), \tag{5.28}$$

$$\text{s.t.} \quad \hat{z}_i(t_k^i \,|\, t_k^i) = z_i(t_k^i), \tag{5.29}$$

$$\dot{\hat{z}}_i(t|\, t_k^i) = f\,(\hat{z}_i(t|\, t_k^i), \hat{u}_i(t|\, t_k^i)), \tag{5.30}$$

$$\hat{u}_i(t|\, t_k^i) \in U_i, \tag{5.31}$$

$$\left\|\hat{p}_{ei}(t|\, t_k^i)\right\| \le r e^{t_k^i + T - t}, \tag{5.32}$$

$$\hat{p}_{ei}(t_k^i + T \,|\, t_k^i) \in \Omega_\varepsilon, \tag{5.33}$$

where r is a constant such that $\Omega_r = \{\hat{p}_{ei} \mid \|\hat{p}_{ei}\| \leq r\} \subseteq \Omega$, and Ω is defined in (5.16). Additionally, $\Omega_\varepsilon = \{\hat{p}_{ei} \mid \|\hat{p}_{ei}\| \leq \varepsilon r\}$, with $0 < \varepsilon < 1$.

The cost function to be optimized is given by:

$$J_i(\hat{p}_{ei}(t_k^i), \hat{u}_i(t_k^i)) = \int_{t_k^i}^{t_k^i+T} L_i(\hat{p}_{ei}(t \mid t_k^i), \hat{u}_i(t \mid t_k^i))\, dt + g_i(\hat{p}_{ei}(t_k^i + T \mid t_k^i)),$$

where the stage cost $L_i(\hat{p}_{ei}(t \mid t_k^i), \hat{u}_i(t \mid t_k^i))$ depends on the state deviation and control input, and is given by (5.12). The terminal cost $g_i(\hat{p}_{ei}(t_k^i + T \mid t_k^i))$ is a function of the terminal state deviation and is designed by (5.11). Equation (5.29) indicates that the initial value of the nominal state is identical to that of the actual state. The robust constraint is represented by (5.32), and it imposes restrictions on the convergence rate of $\hat{p}_{ei}$. The terminal constraint is denoted by (5.33).

The optimization problem formulated by (5.28)–(5.33) is solved over the interval $[t_k^i, t_k^i + T]$, yielding the optimal control input $\hat{u}_i^*(t \mid t_k^i)$ for vehicle i, where the asterisk denotes the optimal solution of the problem. The optimal control input over the interval $[t_k^i, t_{k+1}^i)$ is implemented to vehicle i, and the optimization problem is solved repeatedly at the subsequent computation times.

5.3 Periodic Event-triggering Mechanism Design

In classical model predictive control algorithms, system states must be continuously sampled and updated while repeatedly solving optimization problems, imposing a significant computational burden on the controller. To alleviate this load, an event-triggered mechanism can be adopted, where the optimization problem is solved only at specific time instants, thereby reducing the frequency of computations.

In continuous-time systems employing event-triggered MPC, triggering is typically initiated when the deviation between the actual and predicted states reaches a predefined threshold. However, such a triggering scheme requires dedicated sensors and processors to perform real-time state monitoring and evaluate the triggering condition, imposing stringent demands on the hardware. Moreover, in multi-agent systems, where information exchange among agents is necessary, achieving efficient communication and ensuring the orderly and concise transmission of information becomes a critical challenge.

In addressing the formation control problem for multi-agent systems under model predictive control, this section introduces a periodic event-triggering mechanism. In this mechanism, all agents operate within a unified time framework: they share the same initial time and follow a common sampling period for discretized updates. At each sampling instant, agents assess whether the triggering condition is met, receive information from their neighbors, and transmit their own information as necessary.

The advantages of employing periodic sampling and communication include the alignment of all agents to a consistent temporal scale, with information exchange and state monitoring occurring only at predetermined times. Sensors are therefore required to transmit data and monitor system states only at these fixed intervals, significantly reducing energy consumption associated with continuous sampling and triggering. Moreover, the periodic nature of state monitoring and information exchange lowers hardware demands, and particular equipment for real-time monitoring and trigger evaluation is no longer necessary.

Compared with conventional event-triggering conditions, where triggering is initiated precisely when the deviation between the actual and predicted states equals a predefined threshold, the proposed triggering cannot be achieved under periodic sampling due to the discontinuity in measurement. As a result, in the proposed triggering, the triggering condition must be modified accordingly. A detailed discussion of the periodic sampling event-triggered mechanism is presented below.

The sequence $\{t_k^i\}$, with $k \in \mathbb{N}$, denotes the triggering time instants of vehicle i. Let σ_{val} represent the desired triggering threshold, expressed by:

$$\sigma_{\text{val}} = \frac{\eta}{a}\left(e^{a\delta} - 1\right),$$

where $0 < \delta < T$. Due to the periodic nature of sampling, triggering based on the condition that the deviation between the actual and predicted states exceeds the desired threshold σ_{val} cannot ensure feasibility and system stability. Therefore, the actual triggering of vehicle i must satisfy the following condition:

$$\|p_{ei}(t) - \hat{p}_{ei}^*(t \mid t_k^i)\| \leq \frac{\eta}{a}(e^{a\delta} - 1). \tag{5.34}$$

When employing the proposed periodic event-triggering mechanism, a predictive estimation approach is proposed in this section to proactively assess the triggering condition in order to satisfy condition (5.34). Under this approach, at each sampling instant, the state deviation at the next sampling time is estimated in advance. If the predicted deviation exceeds the desired triggering threshold, it indicates that the next sampling would be unsafe, and the current sampling instant is treated as the triggering time instant. Conversely, if the predicted deviation at the next sampling instant is smaller than the desired threshold, it indicates that the next sampling is safe, and periodic sampling proceeds.

For periodic event-triggering based on the predictive estimation, the sampling period is γT, where $0 < \gamma < \frac{\delta}{T}$. If the current triggering instant of vehicle i is t_k^i, then the next triggering instant t_{k+1}^i can be calculated by:

$$t_{k+1}^i = \min\{t \mid t \in D_{t_k^i}, \left\|p_{ei}(t) - \hat{p}_{ei}^*(t \mid t_k^i)\right\| > \sigma_{\text{safe}}\}, \tag{5.35}$$

where

$$D_{t_k^i} = \left\{ t_k^i + m\gamma T \mid 0 < m \leq \left\lfloor \frac{1}{\gamma} \right\rfloor,\ m \in \mathbb{N} \right\}$$

denotes the set of periodic sampling time instants; the number of sampling intervals between two successive triggering events is denoted by m; the safety threshold for predictive evaluation is calculated by:

$$\sigma_{\text{safe}} = \frac{\eta}{a}\left(e^{a(\delta-\gamma T)} - 1\right)$$

the actual state at sampling time t is $p_{ei}(t)$, and the optimal predicted state at time t, obtained by solving the optimization problem at t_k^i, is denoted by $\hat{p}_{ei}^*(t \mid t_k^i)$.

Since a periodic sampling event-triggered scheme is adopted, an infinite number of triggers within a finite time interval does not occur, thereby effectively avoiding Zeno behavior. Moreover, because the prediction horizon is T, the predicted state $\hat{p}_{ei}^*(t \mid t_k^i)$ beyond this horizon becomes unknown. As a result, during periodic sampling, the number of sampling instants m between two triggering events must be constrained so that it does not exceed $\lfloor 1/\gamma \rfloor$, ensuring that $\hat{p}_{ei}^*(t \mid t_k^i)$ always exists.

The analysis of the triggering time (5.35), derived from the predictive estimation method, is presented below. Suppose that at the sampling instant t_1, condition (5.35) is not satisfied, i.e.,

$$\left\| p_{ei}(t_1) - \hat{p}_{ei}^*(t_1 \mid t_k^i) \right\| \leq \sigma_{\text{safe}}.$$

In this case, the deviation at the next sampling instant $t_1 + \gamma T$ is estimated in advance. Applying the Lipschitz continuity assumption and the Gronwall–Bellman inequality yields:

$$\begin{aligned}
&\left\| p_{ei}(t_1 + \gamma T) - \hat{p}_{ei}^*(t_1 + \gamma T \mid t_k^i) \right\| \\
&\leq \Big\| p_{ei}(t_1) - \hat{p}_{ei}^*(t_1 \mid t_k^i) + \int_{t_1}^{t_1+\gamma T} (f(p_{ei}(s), \hat{u}^*(s|t_k^i)) + n_i(s) \\
&\quad - f(\hat{p}_{ei}^*(s|t_k^i), \hat{u}^*(s|t_k^i)))ds \Big\| \\
&\leq \left\| p_{ei}(t_1) - \hat{p}_{ei}^*(t_1 \mid t_k^i) \right\| + \int_{t_1}^{t_1+\gamma T} \left(a \left\| p_{ei}(s) - \hat{p}_{ei}^*(s \mid t_k^i) \right\| + \eta \right) ds \\
&\leq e^{\gamma T} \frac{\eta}{a}\left(e^{a(\delta-\gamma T)} - 1\right) + \frac{\eta}{a}\left(e^{a\gamma T} - 1\right) \\
&\leq \frac{\eta}{a}\left(e^{a\delta} - 1\right).
\end{aligned} \tag{5.36}$$

It follows from (5.36) and the desired triggering threshold $\sigma_{\text{val}} = \frac{\eta}{a}\left(e^{a\delta} - 1\right)$, that the deviation at the sampling instant $t_1 + \gamma T$ satisfies:

$$\left\| p_{ei}(t_1 + \gamma T) - \hat{p}^*_{ei}(t_1 + \gamma T \mid t^i_k) \right\| \leq \sigma_{\text{val}},$$

indicating that the deviation at the next sampling instant $t_1 + \gamma T$ does not exceed the desired triggering threshold, and therefore the next sampling is safe and can proceed accordingly.

Assume that the sampling instant t_2 is the earliest time that satisfies condition (5.35). Similar to (5.36), it can be concluded that the state deviation at time $t_2 + \gamma T$ may exceed the desired triggering threshold σ_{val}. This indicates that the instant $t_2 + \gamma T$ is unsafe for continued sampling, and therefore t_2 should be treated as the new triggering time.

The comparison between the safety predictive threshold σ_{safe} and the desired triggering threshold σ_{val} reveals that, since

$$\frac{\eta}{a}(e^{a(\delta-\gamma T)} - 1) < \frac{\eta}{a}(e^{a\delta} - 1),$$

the safety threshold is always smaller than the desired threshold. As a safety margin, σ_{safe} plays a critical role in the triggering decision during periodic sampling: it determines whether a new sampling instant is warranted.

If the state deviation at the current sampling instant is less than σ_{safe}, the predictive estimation method ensures that the deviation at the next sampling instant will remain below σ_{val}. In this case, the next sampling is safe, and normal sampling can proceed. Conversely, if the current state deviation exceeds σ_{safe}, then there exists a risk that the deviation at the next sampling instant may surpass σ_{val}, thereby jeopardizing system stability. Under such circumstances, continued waiting would be unsafe, and the current instant should be treated as a new triggering moment.

Based on the preceding discussion, it follows that under the proposed event-triggering mechanism with predictive estimation and periodic sampling, the deviation between the actual and predicted states at each triggering instant does not exceed the prescribed threshold σ_{val}. This result is formally established in Lemma 5.2.

Lemma 5.2 *Suppose that the previous triggering instant of vehicle i is denoted by t^i_k. Under the event-triggered mechanism based on predictive estimation with periodic sampling as specified in (5.35), the deviation between the actual and predicted states at the next triggering instant t^i_{k+1} will be less than the desired triggering threshold σ_{val}, i.e.,*

$$\left\| p_{ei}(t^i_{k+1}) - \hat{p}^*_{ei}(t^i_{k+1} \mid t^i_k) \right\| \leq \sigma_{\text{val}}. \tag{5.37}$$

Proof: Under the event-triggered mechanism based on predictive estimation with periodic sampling, if t_{k+1}^i is the earliest sampling instant satisfying condition (5.35), then according to the triggering criterion (5.35), it follows that the preceding sampling instant $t_{k+1}^i - \gamma T$ satisfies:

$$\left\|p_{ei}(t_{k+1}^i - \gamma T) - \hat{p}_{ei}^*(t_{k+1}^i - \gamma T \middle| t_k^i)\right\| \leq \sigma_{\text{safe}}. \tag{5.38}$$

By applying the Lipschitz assumption and the Gronwall–Bellman inequality, it can be shown that the deviation between the actual and predicted states at time t_{k+1}^i satisfies:

$$\begin{aligned}
&\left\|p_{ei}(t_{k+1}^i) - \hat{p}_{ei}^*(t_{k+1}^i \middle| t_k^i)\right\| \\
&\leq \left\|p_{ei}(t_{k+1}^i - \gamma T) - \hat{p}_{ei}^*(t_{k+1}^i - \gamma T \middle| t_k^i)\right\| + \int_{t_{k+1}^i - \gamma T}^{t_{k+1}^i} \left(a \left\|p_{ei}(s) - \hat{p}_{ei}^*(s \middle| t_k^i)\right\| + \eta\right) ds \\
&\leq e^{\gamma T} \left\|p_{ei}(t_{k+1}^i - \gamma T) - \hat{p}_{ei}^*(t_{k+1}^i - \gamma T \middle| t_k^i)\right\| + \frac{\eta}{a}\left(e^{a\gamma T} - 1\right).
\end{aligned} \tag{5.39}$$

Substituting (5.38) into (5.39) yields:

$$\begin{aligned}
\left\|p_{ei}(t_{k+1}^i) - \hat{p}_{ei}^*(t_{k+1}^i \middle| t_k^i)\right\| &\leq e^{\gamma T}\sigma_{\text{safe}} + \frac{\eta}{a}(e^{a\gamma T} - 1) \\
&\leq e^{\gamma T}\frac{\eta}{a}(e^{a(\delta - \gamma T)} - 1) + \frac{\eta}{a}(e^{a\gamma T} - 1) \\
&\leq \frac{\eta}{a}(e^{a\delta} - 1).
\end{aligned} \tag{5.40}$$

Based on the definition of the desired triggering threshold $\sigma_{\text{val}} = \frac{\eta}{a}(e^{a\delta} - 1)$, and in conjunction with (5.40), it follows that:

$$\left\|p_{ei}(t_{k+1}^i) - \hat{p}_{ei}^*(t_{k+1}^i \middle| t_k^i)\right\| \leq \sigma_{\text{val}}, \tag{5.41}$$

indicating that, at the next triggering instant t_{k+1}^i, the deviation between the actual and predicted states is less than the desired triggering threshold σ_{val}. □

Based on the above analysis, the sampling process between two successive triggering instants under the periodic event-triggered mechanism can be illustrated graphically, as shown in Figure 5.2.

With the event-triggered mechanism based on predictive estimation within a DMPC framework, a control algorithm can be designed for each vehicle i. At the initial time, each vehicle is required to generate a presumed prediction trajectory in advance. This initial trajectory may be taken as constant, using the vehicle's own initial state as its value. After generating the presumed trajectory, each vehicle transmits it to its neighbors. With access to its own state and the received trajectories from neighboring agents, each vehicle can then solve the associated optimization problem at the initial time, thereby completing the initialization of the algorithm.

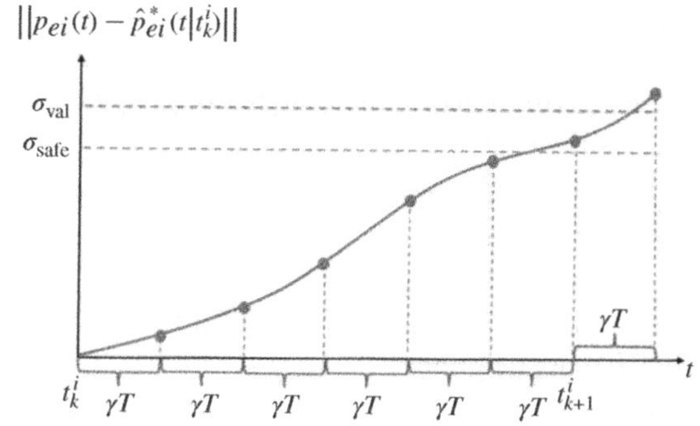

Figure 5.2 Event-triggered mechanism with periodic sampling.

After obtaining the optimal control input sequence by solving the optimization problem using the initialization data, each vehicle continuously applies the control inputs to itself. Subsequently, all vehicles operate on a unified timescale and perform periodic sampling at uniform intervals. After each sampling instance, each vehicle evaluates the triggering condition.

If the condition is satisfied, the vehicle resolves the optimization problem to generate a new control input sequence, computes the corresponding presumed prediction trajectory, and transmits this trajectory to its neighboring vehicles. If the condition is not met, the vehicle continues applying the previously computed control input sequence.

Upon receiving the presumed prediction trajectory, neighboring vehicles update their prior trajectories. Once their own triggering condition is met, they utilize the updated presumed trajectory along with their current state information to solve their respective optimization problems and derive new control input sequences.

Based on the above analysis, the DMPC algorithm under the periodic event-triggered framework can be formulated as Algorithm 5.

In Algorithm 5, all information transmission and state measurement are performed at fixed intervals of γT. Provided that the communication bandwidth is sufficient, this enables efficient data exchange. The proposed algorithm does not require continuous monitoring of state information by sensors, nor does it necessitate continuous evaluation of the triggering condition by dedicated processors, thereby alleviating the demand on specialized hardware performance. Moreover, periodic sampling and data transmission help reduce energy consumption of sensors and processors, which is particularly beneficial for extending battery life in outdoor devices.

Algorithm 5 Event-triggering Distributed MPC Based on Periodic Sampling

1: At the initial time t_0, initialize the information of vehicle i by setting $t_0^i = t_0$ and $\hat{z}_i\left(t_0^i \mid t_0^i\right) = z_i\left(t_0^i\right)$. Define the presumed prediction trajectory as $\overline{p}_{ei}\left(t \mid t_0^i\right) = p_{ei}\left(t_0^i \mid t_0^i\right)$, and transmit $\overline{p}_{ei}\left(t \mid t_0^i\right)$ to all neighboring vehicles $j \in \Psi_i$ while receiving the trajectories $\overline{p}_{ej}\left(t \mid t_0^i\right)$ from them. Set $k = 0$.

2: Vehicle i solves the optimization problem formulated by (5.28)–(5.33) to obtain the optimal control input sequence $\hat{u}_i^*\left(t \mid t_k^i\right)$ for $t \in [t_k^i, t_k^i + T]$, and transmits its presumed prediction trajectory to its neighbors. Set $m = 1$.

3: Apply the control input $\hat{u}_i^*\left(t \mid t_k^i\right)$ to vehicle i over the interval $t \in [t_k^i + (m-1)\gamma T,\ t_k^i + m\gamma T]$.

4: At time $t_k^i + m\gamma T$, perform a state measurement for vehicle i, receive the presumed prediction trajectory information transmitted by neighbor j, and check the triggering condition given in (5.35).

5: If the triggering condition (5.35) is not satisfied, set $m = m + 1$ and return to Step 3. If the triggering condition (5.35) is satisfied, record the current time as the triggering instant, receive the presumed prediction trajectory information from neighboring vehicles, set $k = k + 1$, and return to Step 2.

5.4 Feasibility and Stability

This section analyzes the feasibility of the optimization problem and the stability of the closed-loop system. The basic approach involves constructing a feasible control input sequence, examining the existence of a solution to the optimization problem under this feasible control policy, and verifying the decay property of the Lyapunov function. Taking advantage of the fact that the performance of any feasible solution cannot outperform the optimal one, the existence of an optimal solution is established, ensuring both the feasibility of the optimization problem and the stability of the closed-loop system.

Since the optimization problem defined in (5.28)–(5.33) within the model predictive control (MPC) framework is solved numerically, a feasibility analysis is required. Feasibility, in this context, refers to ensuring that the optimization problem remains solvable at each decision instant—that is, the existence of admissible control inputs that satisfy the control constraints, robust state constraints, and terminal conditions. The core idea of the feasibility analysis is to assume the initial feasibility of the optimization problem, then construct a candidate control sequence based on this solution and demonstrate that all constraints are satisfied at the subsequent optimization step. By invoking mathematical induction, this reasoning can be extended to show that the optimization problem remains feasible at all future decision instants. A detailed feasibility analysis is presented below.

Lemma 5.3 *For vehicle i, suppose that the optimization problem (5.28)–(5.33) is feasible at time t_k^i, i.e., there exists an optimal control input $\hat{u}_i^*(t|\, t_k^i)$ over the prediction horizon $[t_k^i, t_k^i + T]$. Apply the periodic sampling event-triggered MPC in Algorithm 5, and implement $\hat{u}_i^*(t|\, t_k^i)$ to the actual system (5.7) over the horizon $[t_k^i, t_{k+1}^i]$. Provided that*

$$\eta \leq (1-\varepsilon)\frac{\text{a} r e^{(\gamma-1)T}}{e^{a\delta}-1}, \quad \min\{k_1, k_2\}\gamma T \geq \ln\frac{1}{\varepsilon}, \tag{5.42}$$

then at time t_{k+1}^i, there exists a feasible control sequence $\tilde{u}_i(t|\, t_{k+1}^i)$ and a corresponding state sequence $\tilde{p}_{ei}(t\,|t_{k+1}^i)$ satisfying $\tilde{p}_{ei}(t_{k+1}^i + T|\, t_{k+1}^i) \in \Omega_\varepsilon$.

Proof: For vehicle i, if the optimization problem (5.28)–(5.33) is feasible at time t_k^i, then a feasible control sequence $\tilde{u}_i(t|\, t_{k+1}^i)$ can be constructed at the subsequent triggering instant t_{k+1}^i by:

$$\tilde{u}_i(t|\, t_{k+1}^i) = \begin{cases} \hat{u}_i^*(t|\, t_k^i), & t \in [t_{k+1}^i, t_k^i + T], \\ \mu_i(\tilde{p}_{ei}(t\,|t_{k+1}^i)), & t \in [t_k^i + T, t_{k+1}^i + T], \end{cases} \tag{5.43}$$

where $\mu_i(\tilde{p}_{ei}(t\,|t_{k+1}^i))$ denotes the auxiliary terminal controller designed according to Lemma 5.1; $\tilde{p}_{ei}(t\,|t_{k+1}^i)$ represents the nominal system state associated with the application of the feasible control input $\tilde{u}_i(t|\, t_{k+1}^i)$, and it satisfies $\tilde{p}_{ei}(t_{k+1}^i\,|t_{k+1}^i) = p_{ei}(t_{k+1}^i)$.

Starting from time t_{k+1}, the feasible control input $\tilde{u}_i(t|\, t_{k+1}^i)$ is applied to the actual vehicle i system. Over the interval $[t_{k+1}^i, t_k^i + T)$, the deviation between $\tilde{p}_{ei}(t\,|t_{k+1}^i)$ and $\hat{p}_{ei}^*(t\,|t_k^i)$ satisfies:

$$\begin{aligned} &\|\tilde{p}_{ei}(t|t_{k+1}^i) - \hat{p}_{ei}^*(t\,|t_k^i)\| \\ &\leq \|p_{ei}(t_{k+1}^i) - \hat{p}_{ei}^*(t_{k+1}\,|t_k) + \int_{t_{k+1}^i}^{t} f(\tilde{p}_{ei}(s\,|t_{k+1}^i), \tilde{u}_i(s\,|t_{k+1}^i)) \\ &-f(\hat{p}_{ei}^*(s\,|t_k^i), \hat{u}_i^*(s\,|t_k^i))\, ds\|. \end{aligned} \tag{5.44}$$

According to the feasible control (5.43), it follows that $\tilde{u}_i(t|\, t_{k+1}^i) = \hat{u}_i^*(t|\, t_k^i)$ holds over the interval $[t_{k+1}^i, t_k^i + T)$. Substituting it into (5.44) and applying norm inequalities and the Lipschitz condition yield that:

$$\begin{aligned} &\|\tilde{p}_{ei}(t|t_{k+1}^i) - \hat{p}_{ei}^*(t\,|t_k^i)\| \leq \|p_{ei}(t_{k+1}^i) - \hat{p}_{ei}^*(t_{k+1}\,|t_k)\| \\ &+ \int_{t_{k+1}^i}^{t} a\left\|\tilde{p}_{ei}(s\,|t_{k+1}^i) - \hat{p}_{ei}^*(s|\, t_k^i)\right\| ds. \end{aligned} \tag{5.45}$$

It follows from Lemma 5.2 that the state deviation at the triggering time t_{k+1}^{i} satisfies:

$$\left\|p_{ei}\left(t_{k+1}^{i}\right)-\hat{p}_{ei}^{*}\left(t_{k+1}^{i}\middle|\,t_{k}^{i}\right)\right\|\leq\sigma_{\mathrm{val}}. \tag{5.46}$$

Substituting (5.46) into (5.45), and applying Gronwall–Bellman inequality yields that:

$$\left\|\tilde{p}_{ei}\left(t|t_{k+1}^{i}\right)-\hat{p}_{ei}^{*}\left(t|t_{k}^{i}\right)\right\|\leq\sigma_{\mathrm{val}}e^{\left(t-t_{k+1}^{i}\right)}. \tag{5.47}$$

Substitute $t = t_k^i + T$ into (5.47), and apply triangular inequality. It holds that:

$$\left\|\tilde{p}_{ei}(t_k^i+T|t_{k+1}^i)\right\|\leq\sigma_{val}e^{(t_k^i+T-t_{k+1}^i)}+\left\|\hat{p}_{ei}^{*}(t_k^i+T\left|t_k^i\right.)\right\|. \tag{5.48}$$

It follows from the feasibility at time t_k^i that:

$$\left\|\hat{p}_{ei}^{*}(t_k^i+T\left|t_k^i\right.)\right\|\leq\varepsilon r. \tag{5.49}$$

Furthermore, if algorithm (5.35) is employed, the sampling period is given by γT, and therefore the interval between two consecutive triggering instants satisfies:

$$t_{k+1}^i-t_k^i\geq\gamma T. \tag{5.50}$$

Substituting (5.49) and (5.50) into (5.48) yields:

$$\left\|\tilde{p}_{ei}(t_k^i+T|t_{k+1}^i)\right\|\leq\sigma_{\mathrm{val}}e^{(1-\gamma)T}+\varepsilon r. \tag{5.51}$$

Substituting the expected triggering threshold $\sigma_{\mathrm{val}}=\frac{\eta}{a}(e^{a\delta}-1)$ and condition (5.42) into (5.51) yields:

$$\left\|\tilde{p}_{ei}(t_k^i+T|t_{k+1}^i)\right\|\leq r. \tag{5.52}$$

Next, it is necessary to show that the feasible state $\tilde{p}_{ei}(t\left|t_{k+1}^i\right.)$ enters the set $\Omega\varepsilon$ over the interval $t\in[t_k^i+T, tk+1^i+T]$. According to (5.52), we have $\tilde{p}_{ei}(t_k^i + T\left|t_{k+1}^i\right.)\in\Omega_r$ at time t_k^i+T. Given that $\Omega_r\subset\Omega$, it follows that $\tilde{p}_{ei}(t_k^i+T\left|t_{k+1}^i\right.)\in\Omega$, implying that vehicle i lies within the terminal set Ω at time t_k^i+T. From (5.43), it can be seen that over the interval $[t_k^i+T, t_{k+1}^i+T]$, vehicle i implements the auxiliary terminal controller $\mu_i(\tilde{p}_{ei}(t\left|t_{k+1}^i\right.))$. By applying Lemma 5.1, the derivative of $\tilde{p}_{ei}(t\left|t_{k+1}^i\right.)$ can be calculated by:

$$\begin{aligned}\frac{d}{dt}\left\|\tilde{p}_{ei}(t\left|t_{k+1}^i\right.)\right\|^2&=2\hat{x}_{ei}\dot{\hat{x}}_{ei}+2\hat{y}_{ei}\dot{\hat{y}}_{ei}=-2(k_1\tilde{x}_{ei}^2+k_2\tilde{y}_{ei}^2)\\&\leq-2\min\{k_1,k_2\}\left\|\bar{p}_{ei}(t|\,t_{k+1})\right\|^2.\end{aligned} \tag{5.53}$$

Over the interval $[t_k^i+T, t_{k+1}^i+T]$, applying the comparison lemma to (5.53) yields:

$$\left\|\tilde{p}_{ei}(t_{k+1}^i+T|\,t_{k+1}^i)\right\|^2\leq\left\|\tilde{p}_{ei}(t_k^i+T|\,t_{k+1}^i)\right\|^2e^{-2\min\{k_1,k_2\}(t_{k+1}^i-t_k^i)}. \tag{5.54}$$

According to the periodic sampling event-triggered mechanism, the interval between two triggering instants satisfies $t^i_{k+1} - t^i_k \geq \gamma T$. Substituting this into (5.54) and considering (5.52) yields that:

$$\left\| \tilde{p}_{ei}(t^i_{k+1} + T \middle| t^i_{k+1}) \right\|^2 \leq r^2 e^{-2\gamma T \min\{k_1, k_2\}}. \tag{5.55}$$

Substituting condition (5.42) into (5.55) and taking the square root of the resulting inequality yields that:

$$\| \tilde{p}_{ei}(t^i_{k+1} + T \middle| t^i_{k+1}) \| \leq \varepsilon r. \tag{5.56}$$

indicating that, at time $t^i_{k+1} + T$, the feasible state $\tilde{p}_{ei}(t^i_{k+1} + T \middle| t^i_{k+1})$ lies within the set Ωε, i.e., $\tilde{p}_{ei}(t^i_{k+1} + T \middle| t^i_{k+1}) \in \Omega_\varepsilon$. □

Lemma 5.4 *For vehicle i, suppose that the optimization problem (5.28)–(5.33) is feasible at time t^i_k with optimal solution $\hat{u}^*_i(t| t^i_k)$. Apply the periodic sampling event-based MPC given by Algorithm 5. Provided that condition (5.42) and*

$$e^{\gamma T} - 1 \geq 1 - \varepsilon \tag{5.57}$$

are satisfied, then at triggering time t^i_{k+1} there exists a feasible control sequence $\tilde{u}_i(t| t^i_{k+1})$ and corresponding state sequence satisfying $\|\tilde{p}_{ei}(t| t^i_{k+1})\| \leq r e^{t^i_{k+1}+T-t}$ over the horizon $t \in [t^i_{k+1}, t^i_{k+1} + T]$.

Proof: For vehicle i, construct a feasible control at time t^i_{k+1} by (5.43).

At start, consider the interval $t \in [t^i_{k+1}, t^i_k + T)$. According to (5.47) in Lemma 5.3, and by applying the triangle inequality, it follows that:

$$\left\| \tilde{p}_{ei}\left(t | t^i_{k+1}\right) \right\| \leq \left\| \hat{p}^*_{ei}\left(t | t^i_k\right) \right\| + \sigma_{\text{val}} e^{\left(t - t^i_{k+1}\right)}. \tag{5.58}$$

Substituting the expected triggering condition $\sigma_{\text{val}} = \frac{\eta}{a}(e^{a\delta} - 1)$ into (5.58), and applying (5.42) yields that:

$$\begin{aligned} \|\tilde{p}_{ei}(t|t^i_{k+1})\| &\leq \|\hat{p}^*_{ei}(t\,|t^i_k)\| + \frac{\eta}{a}(e^{a\delta} - 1)e^{(t-t^i_{k+1})} \\ &\leq \|\hat{p}^*_{ei}(t\,|t^i_k)\| + (1-\varepsilon) r e^{(t-t^i_{k+1}+\gamma T - T)}. \end{aligned} \tag{5.59}$$

Under the periodic-sampling triggering mechanism, the inter-trigger interval satisfies $t^i_{k+1} - t^i_k \geq \gamma T$. Substituting this into (5.59) yields:

$$\begin{aligned} \|\tilde{p}_{ei}(t|t^i_{k+1})\| &\leq \|\hat{p}^*_{ei}(t\,|t^i_k)\| + (1-\varepsilon) r e^{(t-t^i_{k+1}+t^i_{k+1}-t^i_k-T)} \\ &\leq \|\hat{p}^*_{ei}(t\,|t^i_k)\| + (1-\varepsilon) r e^{t-(t^i_k+T)}. \end{aligned} \tag{5.60}$$

Since the interval $t \in [t_{k+1}^i, t_k^i + T)$ is considered, it follows that $t - (t_k^i + T) < 0$. Substituting this into (5.60) yields:

$$\|\tilde{p}_{ei}(t|t_{k+1}^i)\| \leq \|\hat{p}_{ei}^*(t\,|t_k^i)\| + (1-\varepsilon)r. \tag{5.61}$$

Since the optimization problem (5.28)–(5.33) admits an optimal solution at time t_k, it follows that $\left\|\hat{p}_{ei}^*(t|\,t_k^i)\right\| \leq re^{t_k^i+T-t}, t \in [t_k^i, t_k^i + T]$ for all $t \in [t_k^i, t_k^i + T]$. Substituting this into (5.61) yields:

$$\left\|\tilde{p}_{ei}(t|\,t_{k+1}^i)\right\| \leq re^{t_k^i+T-t} + r - \varepsilon r. \tag{5.62}$$

Substituting condition (5.57) into (5.62) yields:

$$\left\|\tilde{p}_{ei}(t|\,t_{k+1}^i)\right\| \leq re^{t_k^i+T-t} + r(e^{\gamma T} - 1). \tag{5.63}$$

To prove $\left\|\tilde{p}_{ei}(t|\,t_{k+1}^i)\right\| \leq re^{t_{k+1}^i+T-t}$, we can construct a function:

$$y(t) = re^{t_{k+1}^i+T-t} - re^{t_k^i+T-t} - r(e^{\gamma T} - 1), \quad t \in [t_{k+1}^i, t_k^i + T). \tag{5.64}$$

According to the periodic-sampling triggering mechanism, the inter-trigger relationship satisfies $t_{k+1}^i = t_k^i + m\gamma T$. Substituting this into (5.64) yields:

$$y(t) = re^{t_k^i+T}(e^{m\gamma T} - 1)e^{-t} - r(e^{\gamma T} - 1). \tag{5.65}$$

Since e^{-t} is monotonically decreasing with respect to t, the function $y(t)$ in (5.65) is decreasing over the interval $t \in [t_{k+1}^i, t_k^i + T)$. By substituting $t = t_k^i + T$ into (5.65), it follows that:

$$\begin{aligned} y(t_k^i + T) &= re^{t_k^i+T}(e^{m\gamma T} - 1)e^{-(t_k^i+T)} - r(e^{\gamma T} - 1) \\ &= r(e^{m\gamma T} - 1) - r(e^{\gamma T} - 1) \\ &= r(e^{m\gamma T} - e^{\gamma T}). \end{aligned} \tag{5.66}$$

Since $m \geq 1$, it follows that $y(t_k^i + T) \leq 0$. Based on the monotonic decrease of the function $y(t)$, it holds that $y(t) \geq 0$ over the interval $t \in [t_{k+1}^i, t_k^i + T)$. Substituting this into (5.64) yields:

$$re^{t_k^i+T-t} + r(e^{\gamma T} - 1) \leq re^{t_{k+1}^i+T-t}. \tag{5.67}$$

Substituting (5.67) into (5.63) yields:

$$\left\|\tilde{p}_{ei}(t|\,t_{k+1}^i)\right\| \leq re^{t_{k+1}^i+T-t}, \quad t \in [t_{k+1}^i, t_k^i + T). \tag{5.68}$$

Consider the interval $t \in [t_k^i + T, t_{k+1}^i + T)$. From (5.52), it follows that $\|\tilde{p}_{ei}(t_k^i + T\,|\,t_{k+1}^i)\| \leq r$. When the auxiliary terminal controller is employed,

Lemma 5.1 guarantees that $\|\tilde{p}_{ei}(t \mid t^i_{k+1})\| \leq r$ holds throughout the interval $t \in [t^i_k + T, t^i_{k+1} + T)$. Since $r \leq re^{t^i_{k+1}+T-t}$ for $t \in [t^i_k + T, t^i_{k+1} + T)$, it can be concluded that:

$$\left\|\tilde{p}_{ei}(t \mid t^i_{k+1})\right\| \leq re^{t^i_{k+1}+T-t}, \quad t \in [t^i_k + T, t^i_{k+1} + T). \tag{5.69}$$

According to (5.68) and (5.69), if the optimization problem admits an optimal solution $\hat{u}^*_i(t \mid t^i_k)$ at time t^i_k, and conditions (5.42) and (5.57) are satisfied, then under Algorithm 5, a feasible control sequence $\tilde{u}_i(t \mid t^i_{k+1})$ constructed at the triggering instant t^i_{k+1} ensures that $\|\tilde{p}_{ei}(t \mid t^i_{k+1})\| \leq re^{t^i_{k+1}+T-t}$ holds. □

Based on Lemmas 5.3 and 5.4, the feasibility of the optimization problem can be derived. The feasibility result is formally stated in Theorem 5.1.

Theorem 5.1 For vehicle i, suppose that the optimization problem (5.28)–(5.33) is feasible at the initial time t_0. Then, if condition (5.42) in Lemma 5.3 and condition (5.57) in Lemma 5.4 are satisfied, the optimization problem remains feasible at all subsequent triggering instants under the proposed event-triggered MPC with periodic sampling described by Algorithm 5.

Proof: Mathematical induction is employed to establish the feasibility of the optimization problem (5.28)–(5.33) at each triggering instant.

For vehicle i, suppose that the optimization problem (5.28)–(5.33) admits a solution at time t^i_k. Then, it is necessary to analyze the existence of a solution at the next triggering instant t^i_{k+1}.

We begin by analyzing the control constraint. The feasible control input at the triggering instant t^i_{k+1}, denoted as $\tilde{u}_i(t \mid t^i_{k+1})$, is constructed according to (5.43). Over the interval $t \in [t^i_{k+1}, t^i_k + T)$, the feasible control input is given by the optimal control signal $\hat{u}_i(t \mid t^i_k)$ obtained at time t^i_k. Since the optimization problem admits a solution at t^i_k, we have $\hat{u}^*_i(t, |, t_k i) \in U_i$ for all $t \in [t^i_{k+1}, t^i_k + T)$, thereby ensuring that the feasible control input satisfies the constraint over this interval. For $t \in [t^i_k + T, t^i_{k+1} + T)$, the feasible control input is given by the auxiliary terminal controller $\mu_i(\tilde{p}_{ei}(t \mid t^i_{k+1}))$, satisfying the control constraint in view of Lemma (5.1). Therefore, the feasible control input satisfies the control constraint (5.31) over the entire interval $t \in [t^i_{k+1}, t^i_{k+1} + T)$.

Next, according to Lemma 5.3, it follows that $\tilde{p}_{ei}(t^i_{k+1} + T \mid t^i_{k+1}) \in \Omega_\varepsilon$, indicating that the terminal constraint (5.33) in the optimization problem is satisfied. In addition, Lemma 5.4 implies that $\|\tilde{p}_{ei}(t \mid t^i_{k+1})\| \leq re^{t^i_{k+1}+T-t}$, ensuring that the contraction constraint (5.32) is fulfilled. Finally, under the periodic sampling-triggered algorithm, both the initial condition constraint (5.29) and the system dynamics constraint (5.30) are satisfied. Therefore, at time t^i_{k+1}, there exists a feasible solution $\tilde{u}_i(t \mid t^i_{k+1})$ such that the optimization problem (5.28)–(5.33) is satisfied.

In summary, the optimization problem (5.28)–(5.33) is recursively feasible. □

The closed-loop stability with Algorithm 5 is given in the following theorem.

Theorem 5.2 For vehicle i, apply the event-triggered MPC with periodic sampling described by Algorithm 5. If all conditions in Theorem 5.1 are satisfied, and the following conditions hold:

$$1)\quad -k_1 + q_{\max} + p_{\max}k_1^2 \leq 0, \quad -k_2 + q_{\max} + p_{\max}k_2^2 \leq 0 \tag{5.70}$$

$$2)\quad \beta > \alpha_1 + \alpha_2 + \alpha_3 + \alpha_4 \tag{5.71}$$

where $q_{\max} = \max\{q_1, q_2\}$, $q_{\min} = \min\{q_1, q_2\}$, $p_{\max} = \max\{p_1, p_2\}$, $h_{\max} = \max\{h_1, h_2\}$ and

$$\begin{aligned}
\beta &= q_{\min}\gamma T\varepsilon^2 r^2,\\
\alpha_1 &= q_{\max}\sigma_{\text{val}}(e^{2(1-\gamma T)} - 1)(\frac{\sigma_{\text{val}}}{2} + r),\\
\alpha_2 &= \sigma_{\text{val}}e^{(1-\gamma)T}\varepsilon r + \frac{\sigma_{\text{val}}^2 e^{2(1-\gamma)T}}{2},\\
\alpha_3 &= \sum_{j\in\Psi_i} h_{\max}(2r\sigma_{\text{val}}e^{T} + 4r^2e^{2T})(1-\gamma)T + 4r^2e^{2T} + 2r\sigma_{\text{val}}e^{T}(e^{(1-\gamma)T} - 1),\\
\alpha_4 &= \sum_{j\in\Psi_i} 2r^2 h_{\max}T,
\end{aligned}$$

then the tracking error converges into the set $\Omega_p = \{p_e \,|\, \|p_e\| \leq \frac{\eta}{a}(e^{a\delta} - 1) + r\}$.

Proof: For vehicle i, the optimal cost function of the optimization problem is used as the Lyapunov function candidate, denoted by $V_i(p_{ei}(t))$. At time t_k^i, it is given by:

$$\begin{aligned}
V_i(p_{ei}(t_k^i)) &= J_i(\hat{p}_{ei}^*(t_k^i), \hat{u}_i^*(t_k^i))\\
&= \int_{t_k^i}^{t_k^i+T} L_i(\hat{p}_{ei}^*(t|\, t_k^i), \hat{u}_i^*(t|\, t_k^i))dt + g_i(\hat{p}_{ei}^*(t_k^i + T|\, t_k^i)).
\end{aligned} \tag{5.72}$$

Then, the difference between the Lyapunov candidate values at the triggering instants t_{k+1}^i and t_k^i satisfies:

$$\begin{aligned}
\Delta V_i &= V_i(p_{ei}(t_{k+1}^i)) - V_i(p_{ei}(t_k^i)) \leq J_i(\tilde{p}_{ei}(t_{k+1}^i), \tilde{u}_i(t_{k+1}^i)) - J_i(\hat{p}_{ei}^*(t_k^i), \hat{u}_i^*(t_k^i))\\
&\leq \Delta V_{i1} + \Delta V_{i2} + \Delta V_{i3} + \Delta V_{ij1} + \Delta V_{ij2} + \Delta V_{ij3},
\end{aligned} \tag{5.73}$$

where

$$\Delta V_{i1} = \int_{t_{k+1}^i}^{t_k^i+T} \left(\left\| \tilde{p}_{ei}\left(t \left| t_{k+1}^i\right.\right) \right\|_Q^2 - \left\| \hat{p}_{ei}^*\left(t \left| t_k^i\right.\right) \right\|_Q^2 \right) dt,$$

$$\Delta V_{i2} = \int_{t_k^i+T}^{t_{k+1}^i+T} \left(\left\| \tilde{p}_{ei}\left(t \left| t_{k+1}^i\right.\right) \right\|_Q^2 + \left\| \tilde{u}_{ei}\left(t \left| t_{k+1}^i\right.\right) \right\|_P^2 \right) dt + \frac{1}{2}\left\| \tilde{p}_{ei}\left(t_{k+1}^i + T \left| t_{k+1}^i\right.\right) \right\|^2,$$

$$-\frac{1}{2}\left\| \hat{p}_{ei}^*\left(t_k^i + T \left| t_k^i\right.\right) \right\|^2$$

$$\Delta V_{i3} = -\int_{t_k^i}^{t_{k+1}^i} \left(\left\| \hat{p}_{ei}^*\left(t \left| t_k^i\right.\right) \right\|_Q^2 + \left\| \hat{u}_{ei}^*\left(t \left| t_k^i\right.\right) \right\|_P^2 \right) dt,$$

$$\Delta V_{ij1} = \sum_{j\in\Psi_i} \int_{t_{k+1}^i}^{t_k^i+T} \left(\left\| \tilde{p}_{ei}\left(t \left| t_{k+1}^i\right.\right) - \overline{p}_{ej}\left(t \left| t_{k+1}^i\right.\right) \right\|_H^2 - \left\| \hat{p}_{ei}^*\left(t \left| t_k^i\right.\right) - \overline{p}_{ej}\left(t \left| t_k^i\right.\right) \right\|_H^2 \right) dt,$$

$$\Delta V_{ij2} = \sum_{j\in\Psi_i} \int_{t_k^i+T}^{t_{k+1}^i+T} \left\| \tilde{p}_{ei}\left(t \left| t_{k+1}^i\right.\right) - \overline{p}_{ej}\left(t \left| t_{k+1}^i\right.\right) \right\|_H^2 dt,$$

$$\Delta V_{ij3} = \sum_{j\in\Psi_i} -\int_{t_k^i}^{t_{k+1}^i} \left\| \hat{p}_{ei}^*\left(t \left| t_k^i\right.\right) - \overline{p}_{ej}\left(t \left| t_k^i\right.\right) \right\|_H^2 dt.$$

If vehicle i is initially located outside the set Ω_r, then for ΔV_{i1}, it can be derived using (5.47) and norm inequalities that:

$$\begin{aligned}
\Delta V_{i1} &\leq \int_{t_{k+1}^i}^{t_k^i+T} \left(\left\| \tilde{p}_{ei}\left(t|t_{k+1}^i\right) - \hat{p}_{ei}^*\left(t|t_k^i\right) \right\|_Q \right) \left(\left\| \tilde{p}_{ei}\left(t|t_{k+1}^i\right) \right\|_Q + \left\| \hat{p}_{ei}^*\left(t|t_k^i\right) \right\|_Q \right) dt \\
&\leq \int_{t_{k+1}^i}^{t_k^i+T} q_{max}\sigma_{val}e^{\left(t-t_{k+1}^i\right)} \left(2\left\| \hat{p}_{ei}^*(t|t_k^i) \right\| + \sigma_{val}e^{\left(t-t_{k+1}^i\right)} \right) dt \\
&\leq \int_{t_{k+1}^i}^{t_k^i+T} q_{max}\sigma_{val}^2 e^{2\left(t-t_{k+1}^i\right)} dt + \int_{t_{k+1}^i}^{t_k^i+T} 2q_{max}\sigma_{val}e^{\left(t-t_{k+1}^i\right)} \left\| \hat{p}_{ei}^*\left(t|t_k^i\right) \right\| dt.
\end{aligned} \tag{5.74}$$

For the first integral term in (5.74), applying the condition $t_{k+1}^i - t_k^i \geq \gamma T$ and performing the integration yield:

$$\int_{t_{k+1}^i}^{t_k^i+T} q_{max}\sigma_{val}^2 e^{2\left(t-t_{k+1}^i\right)} dt \leq \frac{q_{max}\sigma_{val}^2}{2} \left(e^{2(1-\gamma)T-1} \right). \tag{5.75}$$

For the second integral term in (5.74), the application of Hölder's inequality yields:

$$
\begin{aligned}
&\int_{t_{k+1}^i}^{t_k^i+T} 2q_{max}\sigma_{val}e^{(t-t_{k+1}^i)}\left\|\hat{p}_{ei}^*\left(t\left|t_k^i\right.\right)\right\|dt \\
&\leq\left(\int_{t_{k+1}^i}^{t_k^i+T}\left(2q_{max}\sigma_{val}e^{(t-t_{k+1}^i)}\right)^2dt\right)^{\frac{1}{2}}\left(\int_{t_{k+1}^i}^{t_k^i+T}\left\|\hat{p}_{ei}^*\left(t\left|t_k^i\right.\right)\right\|^2dt\right)^{\frac{1}{2}} \\
&\leq_{max}\sigma_{val}r(e^{2(1-\gamma T)}-1).
\end{aligned}
\tag{5.76}
$$

Substituting (5.75) and (5.76) into (5.74) yields:

$$
\Delta V_{i1}\leq q_{max}\sigma_{val}\left(e^{2(1-\gamma T)}-1\right)\left(\frac{\sigma_{val}}{2}+r\right).
\tag{5.77}
$$

It follows from Lemma 5.1 that, over the interval $[t_k^i+T, t_{k+1}^i+T)$, it holds that:

$$
\begin{aligned}
&\frac{d}{dt}\left(\frac{1}{2}\left\|\tilde{p}_{ei}(t\left|t_{k+1}^i\right.)\right\|^2\right)+\left\|\tilde{p}_{ei}(t|\,t_{k+1}^i)\right\|_Q^2+\left\|\tilde{u}_{ei}(t|\,t_{k+1}^i)\right\|_P^2 \\
&\leq -k_1\tilde{x}_{ei}^2-k_2\tilde{y}_{ei}^2+q_{\max}(\tilde{x}_{ei}^2+\tilde{y}_{ei}^2)+p_{\max}(k_1^2\tilde{x}_{ei}^2+k_2^2\tilde{y}_{ei}^2) \\
&\leq(-k_1+q_{\max}+p_{\max}k_1^2)\tilde{x}_{ei}^2+(-k_2+q_{\max}+p_{\max}k_2^2)\tilde{y}_{ei}^2.
\end{aligned}
\tag{5.78}
$$

Substituting (5.70) into (5.78) yields:

$$
\frac{d}{dt}\left(\frac{1}{2}\left\|\tilde{p}_{ei}(t\left|t_{k+1}^i\right.)\right\|^2\right)+\left\|\tilde{p}_{ei}(t|\,t_{k+1}^i)\right\|_Q^2+\left\|\tilde{u}_{ei}(t|\,t_{k+1}^i)\right\|_P^2\leq 0.
\tag{5.79}
$$

Integrating (5.79) over the interval $[t_k^i+T, t_{k+1}^i+T)$ yields:

$$
\begin{aligned}
&\int_{t_k^i+T}^{t_{k+1}^i+T}(\left\|\tilde{p}_{ei}(t|\,t_{k+1}^i)\right\|_Q^2+\left\|\tilde{u}_{ei}(t|\,t_{k+1}^i)\right\|_P^2)dt \\
&\leq\frac{1}{2}\left\|\tilde{p}_{ei}(t_k^i+T\left|t_{k+1}^i\right.)\right\|^2-\frac{1}{2}\left\|\tilde{p}_{ei}(t_{k+1}^i+T\left|t_{k+1}^i\right.)\right\|^2.
\end{aligned}
\tag{5.80}
$$

Substituting (5.80) into ΔV_{i2} yields:

$$
\begin{aligned}
\Delta V_{i2}&\leq\frac{1}{2}\left\|\tilde{p}_{ei}\left(t_k^i+T\left|t_{k+1}^i\right.\right)\right\|^2-\frac{1}{2}\left\|\hat{p}_{ei}^*\left(t_k^i+T\left|t_k^i\right.\right)\right\|^2 \\
&\leq\frac{1}{2}\left\|\tilde{p}_{ei}\left(t_k^i+T\left|t_{k+1}^i\right.\right)-\hat{p}_{ei}^*\left(t_k^i+T\left|t_k^i\right.\right)\right\|\left(\left\|\tilde{p}_{ei}\left(t_k^i+T\left|t_{k+1}^i\right.\right)\right\|\right. \\
&\quad\left.+\left\|\hat{p}_{ei}^*\left(t_k^i+T\left|t_k^i\right.\right)\right\|\right. \\
&\leq\sigma_{val}e^{(1-\gamma)T}\varepsilon r+\frac{\sigma_{val}^2e^{2(1-\gamma)T}}{2}.
\end{aligned}
\tag{5.81}
$$

For ΔV_{i3}, it can be calculated that:

$$\Delta V_{i3} \leq -\int_{t_k^i}^{t_{k+1}^i} \left\| \hat{p}_{ei}^*(t\,|t_k^i) \right\|_Q^2 dt \leq -q_{\min}\gamma T\varepsilon^2 r^2. \tag{5.82}$$

For ΔV_{ij1}, it can be calculated by:

$$\begin{aligned}
\Delta V_{ij1} \leq & \sum_{j\in\Psi_i} \int_{t_{k+1}^i}^{t_k^i+T} \|(\tilde{p}_{ei}(t|t_{k+1}^i) - \overline{p}_{ej}(t|t_{k+1}^i)) - (\hat{p}_{ei}^*(t|t_k^i) - \overline{p}_{ej}(t|t_k^i))\|_H \\
& (\|\tilde{p}_{ei}(t|t_{k+1}^i) - \overline{p}_{ej}(t|t_{k+1}^i)\|_H + \|\hat{p}_{ei}^*(t|t_k^i) - \overline{p}_{ej}(t|t_k^i)\|_H)dt \\
\leq & \sum_{j\in\Psi_i} h_{\max} \int_{t_{k+1}^i}^{t_k^i+T} (\|\tilde{p}_{ei}(t|t_{k+1}^i) - \hat{p}_{ei}^*(t|t_k^i)\| + \|\overline{p}_{ej}(t|t_k^i)\| + \|\overline{p}_{ej}(t|t_{k+1}^i)\|) \\
& (\|\tilde{p}_{ei}(t|t_{k+1}^i)\| + \|\overline{p}_{ej}(t|t_{k+1}^i)\| + \|\hat{p}_{ei}^*(t|t_k^i)\| + \|\overline{p}_{ej}(t|t_k^i)\|)dt \\
\leq & \sum_{j\in\Psi_i} h_{\max}(2r\sigma_{\text{val}}e^T + 4r^2e^{2T})(1-\gamma)T + 4r^2e^{2T} + 2r\sigma_{\text{val}}e^T(e^{(1-\gamma)T} - 1).
\end{aligned} \tag{5.83}$$

Since $\left\|\tilde{p}_{ei}(t \mid t_{k+1}^i)\right\| \leq r$ and $\left\|\bar{p}_{ej}(t \mid t_{k+1}^i)\right\| \leq r$ hold over the interval $[t_k^i + T, t_{k+1}^i + T)$, it follows that ΔV_{ij2} satisfies:

$$\begin{aligned}
\Delta V_{ij2} &\leq \sum_{j\in\Psi_i} h_{\max} \int_{t_k^i+T}^{t_{k+1}^i+T} \left\|\tilde{p}_{ei}(t\,|t_{k+1}^i)\right\|^2 + \left\|\overline{p}_{ej}(t\,|t_{k+1}^i)\right\|^2 dt \\
&\leq \sum_{j\in\Psi_i} 2r^2 h_{\max} T.
\end{aligned} \tag{5.84}$$

For ΔV_{ij3}, it can be calculated by:

$$\Delta V_{ij3} = \sum_{j\in\Psi_i} \int_{t_k^i}^{t_{k+1}^i} -\left\|\hat{p}_{ei}^*(t\,|t_k^i) - \overline{p}_{ej}(t\,|t_k)\right\|_H^2 dt \leq 0. \tag{5.85}$$

It then follows from (5.77) and (5.81)–(5.85), as well as condition (5.71) that:

$$\Delta V_i \leq \alpha_1 + \alpha_2 + \alpha_3 + \alpha_4 - \beta < 0, \tag{5.86}$$

implying that the Lyapunov function decreases over time, such that $p_{ei}(t)$ enters the set Ω_r ultimately.

In another aspect, if $p_{ei}(t_k)$ lies within the set Ω_r at time t_k, then, since Ω_r is an invariant set for the nominal system, the predicted trajectory $\hat{p}_{ei}(t \mid t_k^i)$ remains within Ω_r. Taking disturbances into account and applying the triggering condition, it can be inferred that the actual state $p_{ei}(t)$ remains within the set $\Omega_p = \left\{p_e | \ \|p_e\| \leq \frac{\eta}{a}(e^{a\delta} - 1) + r\right\}$. □

5.5 A Simulation Example

This section presents simulation results to validate the proposed event-triggered MPC with periodic sampling. Suppose that three mobile robots to perform circular motion in a fixed triangular formation. The virtual leader follows a circular trajectory with a linear velocity of $v_r = 2$ m/s, an angular velocity of $\omega_r = 0.5$ rad/s, and an initial state of $z_r(0) = [0, -4, 0]^T$.

Assume that each follower vehicle has an identical kinematic model, and its control input is subject to the constraint:

$$U_i = \{\, u_i(t) \mid |v_i(t)| \le 3 \text{ m/s},\ |\omega_i(t)| \le 1.5 \text{ rad/s} \,\}, \quad i = 1, 2, 3.$$

The disturbance range is given by:

$$\{\, (n_{xi}(t),\ n_{yi}(t)) \mid n_{xi}(t) \le 0.001,\ n_{yi}(t) \le 0.001 \,\},$$

and in simulation, disturbances are generated randomly within this upper bound. The desired positions of the three follower vehicles relative to the virtual leader, expressed in the virtual leader's local coordinate frame, are given by:

$$p_{d_1} = [0.5,\ 0.1]^T, \quad p_{d_2} = [-0.5,\ 0.5]^T, \quad p_{d_3} = [-0.5,\ -0.5]^T.$$

The initial states of the follower vehicles are given by:

$$z_1(0) = [-0.6,\ -4,\ 0]^T, \quad z_2(0) = [-0.6,\ -3.7,\ 0]^T, \quad z_3(0) = [-0.7,\ -4.3,\ 0]^T.$$

Each vehicle has a head length of $l = 0.5$ m, and the prediction horizon is set to $T = 1.2$ s. The desired event-triggering threshold parameter is specified by $\delta = 0.6$, and the sampling interval is chosen by $\gamma = \frac{1}{6}$. Design parameters are assigned to $r = 0.8$ and $\varepsilon = 0.02$. The weight matrix coefficients are chosen by:

$$Q = \text{diag}\{0.4,\ 0.4\}, \quad P = \text{diag}\{0.1,\ 0.1\}, \quad H = \text{diag}\{0.6,\ 0.6\}.$$

The simulation is conducted using the event-triggered MPC with periodic intermittent sampling presented in Algorithm 5, and the resulting vehicle trajectories are illustrated in Figure 5.3. Analysis of this figure reveals that the three vehicles do not initially form a triangular formation; however, as the simulation progresses, they gradually converge into the desired triangular formation and execute circular motion as specified.

The controls implemented on the vehicles are depicted in Figures 5.4 and 5.5, where it can be seen that the controls always satisfy their constraints.

The trajectories of the tracking error and the error norm are illustrated in Figures 5.6–5.8. Analysis of the tracking error plots and corresponding numerical data indicates that, under the influence of disturbances, the tracking error of each vehicle eventually converges to a neighborhood around the origin and exhibits oscillatory behavior.

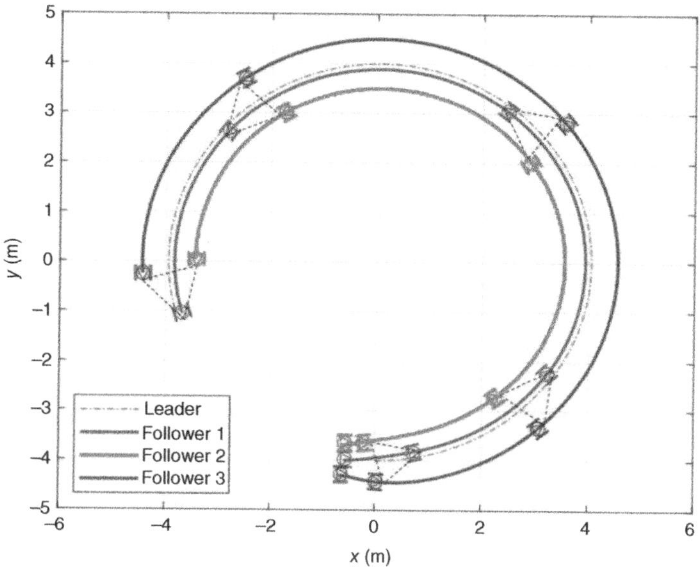

Figure 5.3 Trajectories of vehicles with event-triggered MPC with periodic intermittent sampling.

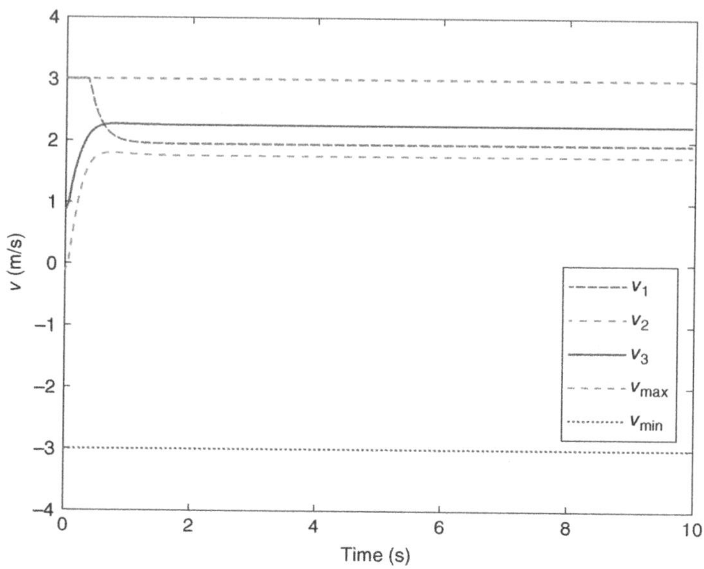

Figure 5.4 Translational velocities of the vehicles.

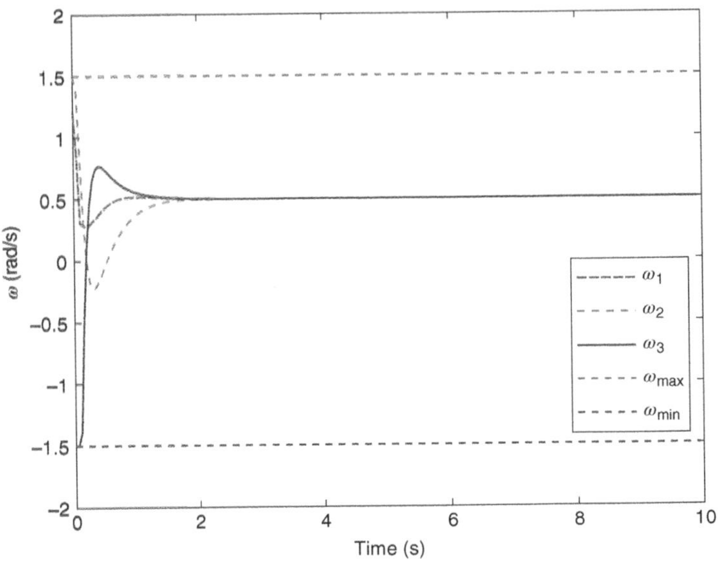

Figure 5.5 Rotational velocities of the vehicles.

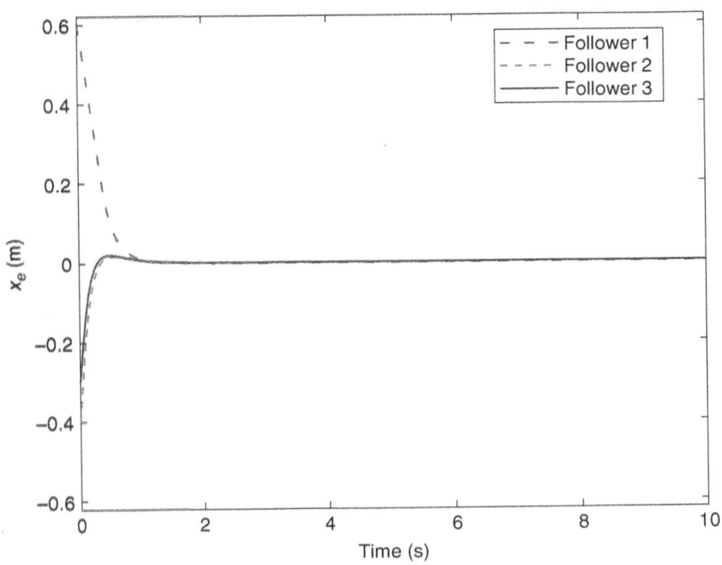

Figure 5.6 Tracking errors in x direction.

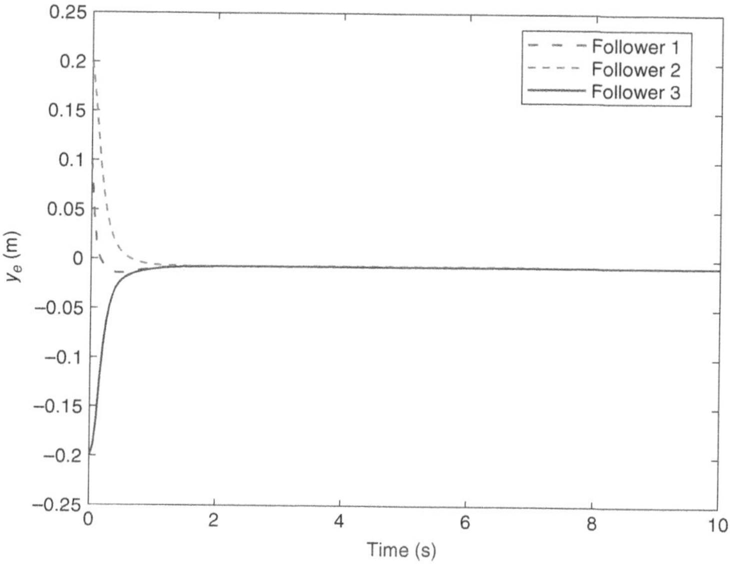

Figure 5.7 Tracking errors in y direction.

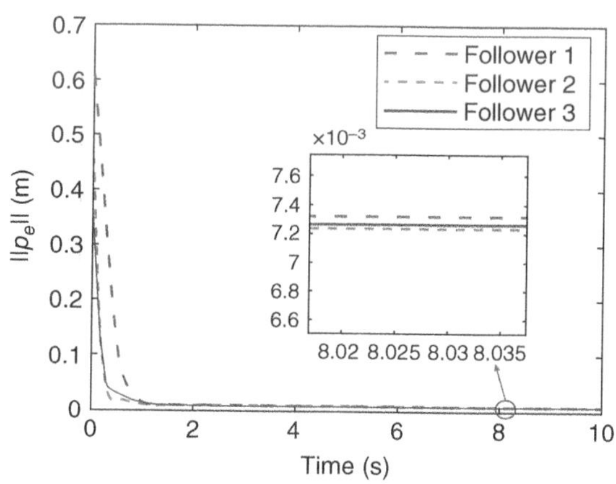

Figure 5.8 Norms of tracking errors.

The sampling instants and optimization instants are depicted in Figure 5.9. Analysis of the figure reveals that, under the event-triggered model predictive control with periodic intermittent sampling, the optimization problem is solved only

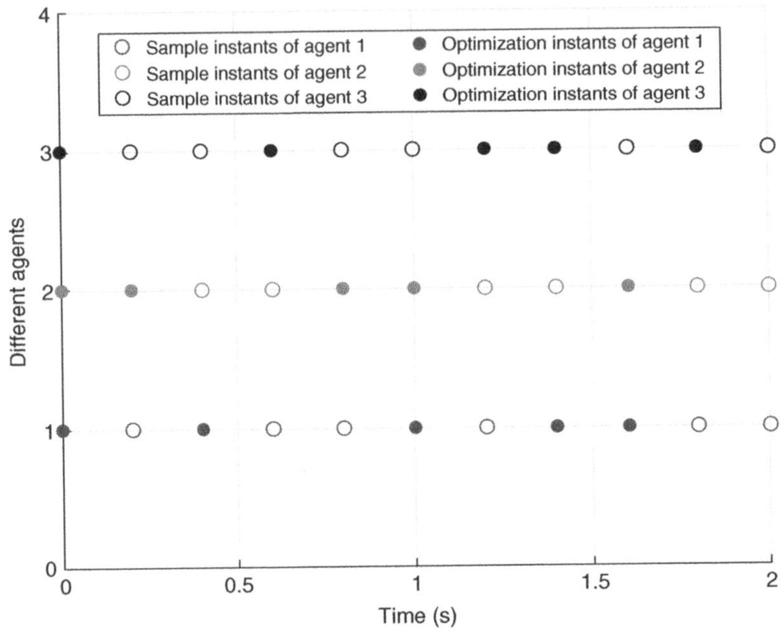

Figure 5.9 Sampling and optimization instants under the event-triggered model predictive control with periodic intermittent sampling.

at specific discrete time instants. Compared to the classical model predictive control method, this algorithm effectively reduces the frequency of both optimization and sampling, thereby alleviating the computational burden on the processor and minimizing the energy consumption resulting from continuous sensor operation.

5.6 Conclusion

This chapter addresses the problem of multi-agent formation control by proposing an event-triggered distributed MPC with periodic intermittent sampling. To mitigate the excessive computational load on processors caused by continuous sampling and optimization in classical MPC methods, an event-triggered mechanism is adopted to reduce the frequency of computation. Within this framework, all agents perform state sampling and information exchange at a constant period, adhering to a unified timescale.

Compared to continuous monitoring and trigger condition evaluation using specialized hardware, the proposed method imposes lower performance requirements on hardware devices. Furthermore, the periodic operation of sensors and processors helps effectively reduce energy consumption. To address the discontinuity in state monitoring caused by periodic sampling, a predictive event-triggered mechanism is employed for designing triggering conditions. This mechanism predicts the state range at the next sampling instant, enabling appropriate triggering to ensure system stability.

For vehicle systems with nonholonomic constraints that cannot be linearized, dedicated terminal state regions and auxiliary terminal controllers are designed. Within the distributed MPC framework, a virtual leader is constructed to serve as a reference agent. Formation maintenance is achieved by incorporating formation cost terms into the objective function, enabling the agents to maintain a fixed geometric configuration during motion.

Rigorous theoretical analysis is conducted to verify the feasibility and stability of the proposed periodically sampled event-triggered multi-agent MPC algorithm. Its effectiveness is further validated through simulation experiments.

6

Concluding Remarks and Future Directions

This book focuses on event-based model predictive control (EMPC) methods. By conducting an in-depth investigation of event-triggering mechanisms, analyzing the characteristics of model predictive control (MPC), studying disturbance-handling strategies, and exploring multi-agent system properties, this book achieves an effective integration of event-based mechanisms and MPC frameworks. From multiple research perspectives, four EMPC schemes are proposed: a two-phase event-triggered MPC method for continuous-time systems, a nonperiodic interval-sampling event-triggered MPC method for discrete-time systems, a composite event-triggered MPC method incorporating disturbance compensation, and a distributed event-triggered MPC method with periodic sampling for multi-agent systems. Through a systematic analysis and integration of the individual features of event-triggering mechanisms and MPC, the proposed approaches effectively reduce the optimization frequency and communication burden inherent in conventional MPC schemes, thereby enhancing their practical applicability. The main contributions can be summarized as follows:

1. For nonlinear continuous-time systems subject to disturbances, an event-based two-phase MPC method is proposed. Compared with existing event-triggered MPC approaches, the proposed method exhibits greater generality, as both periodic event-triggered MPC and self-triggered MPC can be regarded as its special cases. The concept of a safety prediction value is introduced to proactively prevent the invalidation of triggering conditions caused by measurement discontinuities. By properly designing the triggering thresholds and sampling intervals, explicit upper and lower bounds of the triggering interval can be obtained, thereby avoiding the Zeno phenomenon. Furthermore, the tunable parameters allow a trade-off between detection cost and computational cost.

2. For constrained nonlinear discrete-time systems subject to disturbances, an event-based nonperiodic interval-sampling MPC method is developed. This approach fully exploits the feedback information of system states at sampling instants, thereby reducing the conservatism inherent in self-triggered MPC schemes, prolonging the triggering intervals, and lowering the frequency of optimization problem solving. In addition, by introducing a prediction horizon updating strategy through a time-domain contraction factor, the proposed method effectively reduces the computational burden of each optimization step.
3. For constrained systems subject to slowly varying disturbances, a composite event-triggered MPC method with disturbance compensation is proposed. The controller is divided into two components: the disturbance compensation controller, which fully exploits the slowly varying nature of disturbances to enable their estimation and proactive compensation, and the MPC, which optimizes control performance through the solution of an optimization problem. Theoretical analyses of feasibility and stability are provided, and simulation results are presented to verify the effectiveness of the proposed approach.
4. For the formation control of multi-agent vehicle systems, a periodic-sampling event-triggered distributed MPC method is proposed. A virtual leader is constructed as a reference center, enabling each agent to track its desired relative position with respect to this virtual leader. By incorporating a formation cost term into each agent's local optimization problem and enabling information exchange among agents, the proposed scheme achieves fixed-formation coordination for the entire multi-agent system. Furthermore, by performing state monitoring and communication in a periodic-sampling manner, the method ensures efficient and concise information transmission, while the event-triggering mechanism reduces the frequency of optimization problem solving, thereby alleviating the computational burden on local processors.

Although this book has conducted several studies on EMPC methods, there remain many aspects that merit further investigation and exploration. Future research directions include:

1. To address multi-agent formation control problems, future work may further explore how to design event-based distributed MPC architectures that can achieve reliable formation control in the presence of communication delays and packet losses among agents. In addition, considering the influence of obstacles along the motion trajectories, it is also worthwhile to investigate event-based distributed MPC schemes capable of adaptive formation reconfiguration.
2. When using the composite event-triggered MPC method with disturbance compensation, the current approach assumes slowly varying disturbances,

which limits its range of applicability. Future research could focus on more general disturbance scenarios, leveraging disturbance observation or modeling to fully exploit disturbance information within EMPC frameworks for more effective disturbance compensation.

3. The current research primarily addresses external disturbances, while investigations into intrinsic system uncertainties remain limited. Future work could focus on designing EMPC methods for systems with parameter uncertainties or time-varying parameters, thereby enhancing the robustness and practical applicability of the approach. In addition, existing theoretical analyses of external disturbances generally consider their worst-case bounds, which leads to conservative results. Future studies could explore the design of stochastic event-triggered MPC schemes for disturbances following certain probability distributions, aiming to achieve improved system robustness.

References

1 Carlos E Garcia, David M Prett, and Manfred Morari. Model predictive control: Theory and practice—a survey. *Automatica*, 25(3):335–348, 1989.

2 Wilhelmus PMH Heemels, Karl Henrik Johansson, and Paulo Tabuada. An introduction to event-triggered and self-triggered control. In *2012 IEEE 51st IEEE Conference on Decision and Control (SDC)*, pages 3270–3285. IEEE, 2012.

3 Panagiotis D Christofides, Riccardo Scattolini, David Munoz De La Pena, and Jinfeng Liu. Distributed model predictive control: A tutorial review and future research directions. *Computers & Chemical Engineering*, 51:21–41, 2013.

4 Riccardo Scattolini. Architectures for distributed and hierarchical model predictive control–a review. *Journal of Process Control*, 19(5):723–731, 2009.

5 Mazen Alamir, Nida Sheibat-Othman, and Sami Othman. Constrained nonlinear predictive control for maximizing production in polymerization processes. *IEEE Transactions on Control Systems Technology*, 15(2):315–323, 2007.

6 Danylo Malyuta, Yue Yu, Purnanand Elango, and Behçet Açıkmeşe. Advances in trajectory optimization for space vehicle control. *Annual Reviews in Control*, 52:282–315, 2021.

7 Ján Drgoňa, Javier Arroyo, Iago Cupeiro Figueroa, David Blum, Krzysztof Arendt, Donghun Kim, Enric Perarnau Ollé, Juraj Oravec, Michael Wetter, Draguna L Vrabie, et al. All you need to know about model predictive control for buildings. *Annual Reviews in Control*, 50:190–232, 2020.

8 David Q Mayne. Model predictive control: Recent developments and future promise. *Automatica*, 50(12):2967–2986, 2014.

9 David Q Mayne, James B Rawlings, Christopher V Rao, and Pierre OM Scokaert. Constrained model predictive control: Stability and optimality. *Automatica*, 36(6):789–814, 2000.

10 Charles R Cutler and Brian L Ramaker. Dynamic matrix control: A computer control algorithm. In *Joint Automatic Control Conference*, volume 17, page 72, 1980.

11 Jacques Richalet, André Rault, JL Testud, and J Papon. Model predictive heuristic control. *Automatica (Journal of IFAC)*, 14(5):413–428, 1978.

12 David W Clarke, Coorous Mohtadi, and P Simon Tuffs. Generalized predictive control—part i. the basic algorithm. *Automatica*, 23(2):137–148, 1987.

13 James B. Rawlings, David Q Mayne, M. Diehl. *Model Predictive Control: Theory, Computation, and Design*, Nob Hill Publishing: Madison, WI, 2017.

14 EF Camacho and C. Bordons Alba. *Model Predictive Control second edition.* Advanced Textbooks in Control and Signal Processing. Springer: London, 2007. ISBN 9780857293985.

15 Lars Grüne, Jürgen Pannek. *Nonlinear Model Predictive Control.* Springer, 2017.

16 W Kwon and A Pearson. On feedback stabilization of time-varying discrete linear systems. *IEEE Transactions on Automatic Control*, 23(3):479–481, 1978.

17 Hong Chen and Frank Allgöwer. A quasi-infinite horizon nonlinear model predictive control scheme with guaranteed stability. *Automatica*, 34(10):1205–1217, 1998.

18 Shuyou Yu, Marcus Reble, Hong Chen, and Frank Allgöwer. Inherent robustness properties of quasi-infinite horizon nonlinear model predictive control. *Automatica*, 50(9):2269–2280, 2014.

19 David Q Mayne and Wilbur Langson. Robustifying model predictive control of constrained linear systems. *Electronics Letters*, 37(23):1422–1423, 2001.

20 Luigi Chisci, J Anthony Rossiter, and Giovanni Zappa. Systems with persistent disturbances: Predictive control with restricted constraints. *Automatica*, 37(7):1019–1028, 2001.

21 Ilya Kolmanovsky and Elmer G Gilbert. Theory and computation of disturbance invariant sets for discrete-time linear systems. *Mathematical Problems in Engineering*, 4(4):317–367, 1998.

22 David Q Mayne, María M Seron, and Saša V Raković. Robust model predictive control of constrained linear systems with bounded disturbances. *Automatica*, 41(2):219–224, 2005.

23 David Q Mayne, SV Raković, Rolf Findeisen, and Frank Allgöwer. Robust output feedback model predictive control of constrained linear systems: Time varying case. *Automatica*, 45(9):2082–2087, 2009.

24 David Q Mayne, Erric C Kerrigan, EJ Van Wyk, and Paola Falugi. Tube-based robust nonlinear model predictive control. *International Journal of Robust and Nonlinear Control*, 21(11):1341–1353, 2011.

25 Saša V Raković, Basil Kouvaritakis, Mark Cannon, and Christos Panos. Fully parameterized tube model predictive control. *International Journal of Robust and Nonlinear Control*, 22(12):1330–1361, 2012.

26 James Fleming, Basil Kouvaritakis, and Mark Cannon. Robust tube MPC for linear systems with multiplicative uncertainty. *IEEE Transactions on Automatic Control*, 60(4):1087–1092, 2014.

27 Saša V Rakovic and Miroslav Baric. Parameterized robust control invariant sets for linear systems: Theoretical advances and computational remarks. *IEEE Transactions on Automatic Control*, 55(7):1599–1614, 2010.
28 Daniel Limón Marruedo, Teodoro Alamo, and Eduardo F Camacho. Input-to-state stable MPC for constrained discrete-time nonlinear systems with bounded additive uncertainties. In *Proceedings of the 41st IEEE Conference on Decision and Control, 2002*, volume 4, pages 4619–4624. IEEE, 2002.
29 Gilberto Pin, Davide M Raimondo, Lalo Magni, and Thomas Parisini. Robust model predictive control of nonlinear systems with bounded and state-dependent uncertainties. *IEEE Transactions on Automatic Control*, 54(7):1681–1687, 2009.
30 Huiping Li and Yang Shi. Robust distributed model predictive control of constrained continuous-time nonlinear systems: A robustness constraint approach. *IEEE Transactions on Automatic Control*, 59(6):1673–1678, 2013.
31 Peter J Campo and Manfred Morari. Robust model predictive control. In *1987 American Control Conference*, pages 1021–1026. IEEE, 1987.
32 Pierre OM Scokaert and David Q Mayne. Min-max feedback model predictive control for constrained linear systems. *IEEE Transactions on Automatic Control*, 43(8):1136–1142, 1998.
33 Yaohui Lu and Yaman Arkun. Quasi-min-max MPC algorithms for LPV systems. *Automatica*, 36(4):527–540, 2000.
34 Mircea Lazar, D Munoz De La Pena, WP Maurice H Heemels, and Teodoro Alamo. On input-to-state stability of min–max nonlinear model predictive control. *Systems & Control Letters*, 57(1):39–48, 2008.
35 Daniel Limón, Teodoro Alamo, Francisco Salas, and Eduardo F Camacho. Input to state stability of min–max MPC controllers for nonlinear systems with bounded uncertainties. *Automatica*, 42(5):797–803, 2006.
36 Davide Martino Raimondo, Daniel Limon, Mircea Lazar, Lalo Magni, and Eduardo Ferná ndez Camacho. Min-max model predictive control of nonlinear systems: A unifying overview on stability. *European Journal of Control*, 15(1):5–21, 2009.
37 Sergio Lucia, Tiago Finkler, and Sebastian Engell. Multi-stage nonlinear model predictive control applied to a semi-batch polymerization reactor under uncertainty. *Journal of Process Control*, 23(9):1306–1319, 2013.
38 Johan Lofberg. Approximations of closed-loop minimax MPC. In *42nd IEEE International Conference on Decision and Control (IEEE Cat. No. 03CH37475)*, volume 2, pages 1438–1442. IEEE, 2003.
39 Jianchen Hu and Baocang Ding. An efficient offline implementation for output feedback min-max MPC. *International Journal of Robust and Nonlinear Control*, 29(2):492–506, 2019.
40 Paul J Goulart, Eric C Kerrigan, and Jan M Maciejowski. Optimization over state feedback policies for robust control with constraints. *Automatica*, 42(4):523–533, 2006.

41 Sankaranarayanan Subramanian, Sergio Lucia, Radoslav Paulen, and Sebastian Engell. Tube-enhanced multi-stage model predictive control for flexible robust control of constrained linear systems with additive and parametric uncertainties. *International Journal of Robust and Nonlinear Control*, 31(9):4458–4487, 2021.

42 Huiping Li and Yang Shi. Networked min–max model predictive control of constrained nonlinear systems with delays and packet dropouts. *International Journal of Control*, 86(4):610–624, 2013.

43 Ridong Zhang, Sheng Wu, Zhixing Cao, Jingyi Lu, and Furong Gao. A systematic min–max optimization design of constrained model predictive tracking control for industrial processes against uncertainty. *IEEE Transactions on Control Systems Technology*, 26(6):2157–2164, 2017.

44 Changxin Liu, Huiping Li, Jian Gao, and Demin Xu. Robust self-triggered min–max model predictive control for discrete-time nonlinear systems. *Automatica*, 89:333–339, 2018.

45 Jun Yang, and Wei Xing Zheng. Offset-free nonlinear MPC for mismatched disturbance attenuation with application to a static var compensator. *IEEE Transactions on Circuits and Systems II: Express Briefs*, 61(1):49–53, 2013.

46 Junxiao Wang, Fengxiang Wang, Zhenbin Zhang, Shihua Li, and José Rodríguez. Design and implementation of disturbance compensation-based enhanced robust finite control set predictive torque control for induction motor systems. *IEEE Transactions on Industrial Informatics*, 13(5):2645–2656, 2017.

47 Pengfei Li, Yu Kang, Yun-Bo Zhao, and Tao Wang. A novel self-triggered MPC scheme for constrained input-affine nonlinear systems. *IEEE Transactions on Circuits and Systems II: Express Briefs*, 68(1):306–310, 2020.

48 Jaehyun Yoo and Karl H Johansson. Event-triggered model predictive control with a statistical learning. *IEEE Transactions on Systems, Man, and Cybernetics: Systems*, 51(4):2571–2581, 2019.

49 Paolo Varutti, Benjamin Kern, Timm Faulwasser, and Rolf Findeisen. Event-based model predictive control for networked control systems. In *Proceedings of the 48th IEEE Conference on Decision and Control (CDC) held jointly with 2009 28th Chinese Control Conference*, pages 567–572. IEEE, 2009.

50 Daniel Lehmann, Erik Henriksson, and Karl H Johansson. Event-triggered model predictive control of discrete-time linear systems subject to disturbances. In *2013 European Control Conference (ECC)*, pages 1156–1161. IEEE, 2013.

51 Huiping Li and Yang Shi. Event-triggered robust model predictive control of continuous-time nonlinear systems. *Automatica*, 50(5):1507–1513, 2014.

52 Kazumune Hashimoto, Shuichi Adachi, and Dimos V Dimarogonas. Event-triggered intermittent sampling for nonlinear model predictive control. *Automatica*, 81:148–155, 2017.

53 Mengzhi Wang, Jian Sun, and Jie Chen. Input-to-state stability of perturbed nonlinear systems with event-triggered receding horizon control scheme. *IEEE Transactions on Industrial Electronics*, 66(8):6393–6403, 2018.

54 Antonella Ferrara, Gian Paolo Incremona, and Lalo Magni. Model-based event-triggered robust MPC/ISM. In *2014 European Control Conference (ECC)*, pages 2931–2936. IEEE, 2014.

55 Alina Eqtami, Dimos V Dimarogonas, and Kostas J Kyriakopoulos. Event-based model predictive control for the cooperation of distributed agents. In *2012 American Control Conference (ACC)*, pages 6473–6478. IEEE, 2012.

56 Alina Eqtami, Dimos V Dimarogonas, and Kostas J Kyriakopoulos. Novel event-triggered strategies for model predictive controllers. In *2011 50th IEEE Conference on Decision and Control and European Control Conference*, pages 3392–3397. IEEE, 2011.

57 Ning He and Dawei Shi. Event-based robust sampled-data model predictive control: A non-monotonic lyapunov function approach. *IEEE Transactions on Circuits and Systems I: Regular Papers*, 62(10):2555–2564, 2015.

58 Dominic Groß and Olaf Stursberg. A cooperative distributed MPC algorithm with event-based communication and parallel optimization. *IEEE Transactions on Control of Network Systems*, 3(3):275–285, 2015.

59 Kazumune Hashimoto, Shuichi Adachi, and Dimos V Dimarogonas. Distributed aperiodic model predictive control for multi-agent systems. *IET Control Theory & Applications*, 9(1):10–20, 2015.

60 Yuanyuan Zou, Xu Su, Shaoyuan Li, Yugang Niu, and Dewei Li. Event-triggered distributed predictive control for asynchronous coordination of multi-agent systems. *Automatica*, 99:92–98, 2019.

61 Zhongqi Sun, Li Dai, Kun Liu, Dimos V Dimarogonas, and Yuanqing Xia. Robust self-triggered MPC with adaptive prediction horizon for perturbed nonlinear systems. *IEEE Transactions on Automatic Control*, 64(11):4780–4787, 2019.

62 Ning He, Dawei Shi, and Tongwen Chen. Self-triggered model predictive control for networked control systems based on first-order hold. *International Journal of Robust and Nonlinear Control*, 28(4):1303–1318, 2018.

63 Kazumune Hashimoto, Shuichi Adachi, and Dimos V Dimarogonas. Self-triggered model predictive control for nonlinear input-affine dynamical systems via adaptive control samples selection. *IEEE Transactions on Automatic Control*, 62(1):177–189, 2016.

64 Erik Henriksson, Daniel E Quevedo, Henrik Sandberg, and Karl Henrik Johansson. Self-triggered model predictive control for network scheduling and control. *IFAC Proceedings Volumes*, 45(15):432–438, 2012.

65 Jingyuan Zhan, Xiang Li, and Zhong-Ping Jiang. Self-triggered robust output feedback model predictive control of constrained linear systems. In *2017 American Control Conference (ACC)*, pages 3066–3071. IEEE, 2017.

66 JDJ Barradas Berglind, TMP Gommans, and WPMH Heemels. Self-triggered MPC for constrained linear systems and quadratic costs. *IFAC Proceedings Volumes*, 45(17):342–348, 2012.

67 Emre Aydiner, Florian D Brunner, WPMH Heemels, and Frank Allgöwer. Robust self-triggered model predictive control for constrained discrete-time lti systems based on homothetic tubes. In *2015 European Control Conference (ECC)*, pages 1587–1593. IEEE, 2015.

68 Jicheng Chen, Qi Sun, and Yang Shi. Stochastic self-triggered MPC for linear constrained systems under additive uncertainty and chance constraints. *Information Sciences*, 459:198–210, 2018.

69 Jie Tao, Lixing Yang, Zheng-Guang Wu, Xiaofeng Wang, and Hongye Su. Lebesgue-approximation model predictive control of nonlinear sampled-data systems. *IEEE Transactions on Automatic Control*, 65(10):4047–4060, 2019.

70 Liang Lu and Jan M Maciejowski. Self-triggered MPC with performance guarantee using relaxed dynamic programming. *Automatica*, 114:108803, 2020.

71 Changxin Liu, Jian Gao, Huiping Li, and Demin Xu. Aperiodic robust model predictive control for constrained continuous-time nonlinear systems: An event-triggered approach. *IEEE Transactions on Cybernetics*, 48(5):1397–1405, 2017.

72 Tao Wang, Yu Kang, Pengfei Li, Yun-Bo Zhao, and Hao Tang. Rolling self-triggered distributed MPC for dynamically coupled nonlinear systems. *Automatica*, 160:111444, 2024.

73 Hassan K Khalil. Nonlinear systems third edition. *Prentice Hall*, 2002.

74 Huahui Xie, Li Dai, Yu Luo, and Yuanqing Xia. Robust MPC for disturbed nonlinear discrete-time systems via a composite self-triggered scheme. *Automatica*, 127:109499, 2021.

75 Zhongqi Sun, Chang Li, Jinhui Zhang, and Yuanqing Xia. Dynamic event-triggered MPC with shrinking prediction horizon and without terminal constraint. *IEEE Transactions on Cybernetics*, 52(11):12140–12149, 2021.

76 John M Holte. Discrete Gronwall lemma and applications. In *MAA-NCS Meeting at the University of North Dakota*, volume 24, pages 1–7, 2009.

77 Rohan Chandra Shekhar. *Variable horizon model predictive control: Robustness and optimality*. PhD thesis, University of Cambridge, 2012.

78 Aoyun Ma, Kun Liu, Qirui Zhang, Tao Liu, and Yuanqing Xia. Event-triggered distributed MPC with variable prediction horizon. *IEEE Transactions on Automatic Control*, 66(10):4873–4880, 2020.

79 MathWorks. MATLAB and simulink documentation. R2020a, MathWorks, 2020. www.mathworks.com/help/releases/R2020a.html.

Index

Printed and bound by CPI Group (UK) Ltd, Croydon, CR0 4YY
19/09/2026
14965102-0004